Ethics in Science

Ethics in Science
Ethical Misconduct in Scientific Research

Second Edition

John G. D'Angelo
Alfred University

CRC Press
Taylor & Francis Group
Boca Raton London New York

CRC Press is an imprint of the
Taylor & Francis Group, an **informa** business

CRC Press
Taylor & Francis Group
6000 Broken Sound Parkway NW, Suite 300
Boca Raton, FL 33487-2742

© 2019 by Taylor & Francis Group, LLC
CRC Press is an imprint of Taylor & Francis Group, an Informa business

No claim to original U.S. Government works

Printed on acid-free paper

International Standard Book Number-13: 978-1-138-03542-3 (Paperback)
International Standard Book Number-13: 978-1-138-39244-1 (Hardback)

Visit the Taylor & Francis Web site at
http://www.taylorandfrancis.com

and the CRC Press Web site at
http://www.crcpress.com

This book is dedicated to all of those who have supported me in everything I've done. Without each of you, I am nothing.

Contents

PART A   Overview of Ethics Violations

PART B   Case Studies

Introduction to the First Edition

Just about every year, we hear about a case of ethical violations in some field of science, in general. None of the fields of science are immune to this betrayal. The reasons for these violations of scientific ethics are manifold, and a discussion of these reasons, along with discussions of what constitutes a violation of scientific ethics and how the scientific community detects them is provided in Chapter 1. Certainly, one can argue that these issues cannot be resolved without instituting formal moral code and that, because nearly every one of us has our own individual *moral* code, such initiations cannot be made. However, for science to survive, we **must** agree on a basic set of appropriate behavior, at least *in principle*. Indeed, an overwhelming majority of active and retired scientists have agreed upon and follow such a basic set of appropriate behavior. Unfortunately, for **any** portion of society, there are a few dissenters. These dissenters vary from those who feel the rules do not apply to them, to those who disagree with the rules, and even all the way to those who actively and consciously try to "beat the system." Perhaps the most troubling however (only because this group can easily be avoided) are the instances where the perpetrator honestly *did not understand* that their actions were a violation of scientific conduct. Other troubling cases (that are more difficult to address) are where the system actually facilitates, rather than inhibiting the violations.

Although this work will consider the potential misconduct toward human or animal research subjects, it should be noted that it is **not** the intended spirit of this work to delve into a discussion of the "right" and "wrong" of issues such as human and animal testing, cloning, and stem cell research though they will receive brief mention. These are deeply personal and, in many aspects, religious issues and I am nobody to tell you how to feel about them. A discussion of these current issues, especially with attention paid to governments' role in providing funding for such research and popular opposition, is provided in Chapter 5. Instead, it is the intended spirit that focuses on improper conduct in scientific research.

In each of the chapters in Part B, various cases involving *potential* misconduct in science are presented. Every attempt has been made to adequately represent *both* sides of the issue when possible. This has not been done as an attempt to *defend* or *justify* the actions of the putative violator, but instead as an attempt to provide maximum insight into the issue so that something can truly be learned from each case. In each of the cases in Part B, the year and location of the offense (or alleged offense) are provided, along with the names and claims of all of the principal players. When one has been reached, the resolution in the case is also provided. This format allows for several advantages. First, it allows the reader to observe how widespread, both in field and in geographic location, misconduct in science has been, demonstrating that nobody is immune. Ethical violations occur in every country, are committed by people of every race, and are committed (at least allegedly) by even the most famous researchers. Second, it allows the reader to see all the sides and factors of the issue. This has the effect of giving a real firsthand look at why they occur. Finally, when possible, the resolution allows the reader to see what happens to the offenders, those negatively impacted by the offense, and the "whistle-blowers" (those who call attention to the offense). Also, some of the cases are presented specifically because they are so muddied and there has not yet been a resolution, with none on the horizon, either. It is the desired outcome that these latter cases, especially, stimulate discussion in the classroom.

If the cases of scientific misconduct where true ignorance is the cause can be reduced, progress will be made. This education is one of my goals here. It is also a major goal of the American Chemical Society (ACS), which has incorporated a section on science ethics into the new third edition of the ACS Style Guide. The examples that follow here attempt to cover the various forms of ethical violations in science. As will be seen, virtually no one, not even Nobel Prize winners, is immune to controversies regarding scientific misconduct. We will also learn that the specific reasons and the methods of ethical violation/controversy differ almost every time. Hopefully, the discussions that result from this book will act to enlighten students to *what sort of behavior* constitutes scientific misconduct. An additional goal is that it shows that science is in large part self-correcting and self-policing, and that it is truly not worth it to try to get away with it.

HOW TO USE THIS BOOK

It is suggested that this book be used the following way: After reading Chapters 1, 2, and 7, the case study can easily begin. I suggest that students write a summary of **no more than one page** of their thoughts on the case, whenever the case study begins. Also, the instructor should not feel bound to only use cases herein. After an instructor-*mediated* group discussion in class, another **one-page** summary about how their minds were changed or supported by the discussion should be completed. This may allow for a course to be built around this concept to help satisfy a public speaking requirement at some universities. In instances where the instructor would like to build a writing requirement into the course, there is room for flexibility to increase the demands on student writing.

AN IMPORTANT NOTE ON CHAPTERS 3–6

Chapter 3 does not formally cover scientific misconduct. It does, however, address the factors that I feel can contribute to scientific misconduct, at least indirectly. Chapters 4–6 cover increasingly the important issues in the science of today. It is for this reason that I have included them in this book, detached from the discussion about misconduct in science. They are provided for those readers interested in their respective topics, though I think that Part B can certainly be used without having read these four final chapters of Part A.

Introduction to the Second Edition

Years after the first edition was written, there still seems to be too many cases of scientific misconduct in the media; truthfully, any nonzero number is too many, but the pace has not perceptively slowed. Notably, since the inauguration of President Donald Trump, science seems (at least to me) to be under more fire than ever, and some of the science-based government agencies, most especially the Environmental Protection Agency, seem to be in the greatest chaos. I have made no attempt to include some of this chaos here. First, it is too far away from scientific misconduct, the actual topic of this book. Second, I feel as if there is just so (too) much to keep up with. So, these topics will by and large not be covered here. This nevertheless articulates an important point I try to touch upon later: science and decisions for it is best left to scientists. Although scientists are no less human than politicians, they are the ones educated in the field and should be the ones calling the scientific shots. Anything less leads to uninformed decisions being made. But none of this is scientific misconduct per se.

In this second edition, I have expanded on topics such as human subjects and animal testing, peer review, and the responsibility to the public that science holds, and I have added a brief chapter about the pharmaceutical industry and a few new cases to the case studies.

HOW TO USE THIS BOOK

It is suggested that this book be used the following way: after reading Chapters 1, 2, and 9, the case study analysis can easily begin. At a minimum, Chapter 1 must be read before an informed discussion can occur. I suggest that students write a summary of **no more than one page** of their thoughts on the case, whenever the case study begins, though an oral presentation also works well in my experience. This may allow for a course to be built around this concept to help satisfy a public speaking requirement at some universities. In instances where the instructor would like to build a writing requirement into the course, there is room for flexibility to increase the demands on student writing.

ONLINE CONTENT

Additional cases will be uploaded to a content area on the publisher's website. See (www.routledge.com/9781138035423) for more details.

About the Author

I've never liked the idea of myself being referred to in the third or second person; therefore, I am writing this part in the first person.

On September 14, 1978, I was born to Josephine and John D'Angelo. I was blessed with an enormously happy childhood, with two brothers and many cousins nearby and friends to play with; rarely was I bored or alone. A third brother, sadly, watches over all of us and I truly hope that one day I will meet him in paradise.

In my junior year in high school (1994–1995), I took chemistry. It was here that the seeds for my career were planted, watered, and fertilized, and for that I thank Joe Dixon sincerely. After graduation, I went to the State University of New York (SUNY) at Stony Brook, with the intention of becoming a high school teacher. However, after taking Organic Chemistry, I was hooked to it and switched gears to become a college teacher. After graduating in the year 2000, I moved on to the University of Connecticut where I received my PhD in Organic Chemistry under Michael B. Smith in 2005 and subsequently did my Post-Doc at The Johns Hopkins University in the laboratory of the late Gary H. Posner.

In August 2007, I joined the faculty of the Chemistry Department at Alfred University. In 2013, I was awarded tenure and promoted to Associate Professor of Chemistry, my current rank. This is the third book I have authored. The first one was the first edition of this book and the second book was titled *Hybrid Retrosynthesis: Organic Synthesis Using Reaxys and Scifinder*, which I co-authored with my PhD advisor. I am currently working on the content for organic chemistry classes with the online textbook publisher, Top Hat, and expect it to be available in the fall of 2018. This second edition would make four books and I am currently planning a fifth, which would be a workbook for scientific ethics training seminars.

Part A

Overview of Ethics Violations

1 Research Misconduct

What Is It, Why Does It Happen, and How Do We Identify When It Happens?

We should start by finding out what the dictionary has to say about the word *ethics*. And so, according to www.Dictionary.com, ethics (as it applies here) is defined in the following ways:

1. A system of moral principles
2. The rules of conduct recognized in respect to a particular class of human actions or a particular group, culture, etc.

Armed with a definition of ethics, we can ask: what does this mean to science? Unfortunately, the answer to this question is not always clear. Although scientific ethics is almost certainly more #2 than #1, definition #1 almost always asserts itself in individuals, including scientists and engineers. Despite how open to interpretation many ethical issues are, the scientific community has by and large agreed upon a standard of behavioral principles which, **to be clear**, the *vast* majority of practicing scientists largely adhere to. As with every walk of life, however, it is the exceptions that receive most of the attention and ruin it for those who are honest. The baffling paradox is that more often than not, they (the offenders) get caught red-handed and red-faced, ruining their careers. If this is the case, why do they do it? This question is explored later in the context of each violation of scientific conduct as the chapter develops.

Another issue that simply must not be ignored and is covered at the appropriate times in this book is what happens to the person who calls out the offending scientist? However, unfortunate the moniker may be, this person is usually referred to as a "whistleblower." Also, the highly important distinction between the triforce of bad ethics, bad science, and genuine errors must be made. We must remember that science often both proves itself right and proves itself wrong. As we'll come to find, this may be due to genuine mistakes, poorly designed experiments or poorly behaving people. Finally, such errors may also happen because of natural scientific progress. All are possible and are discussed in this book.

WHAT CONSTITUTES SCIENTIFIC MISCONDUCT?

Before embarking on a discussion of why ethical violations in science occur and before ultimately performing a case study of ethical violations in science, we must first **identify** just what scientific misconduct is. This is at least in part because the motivations for doing it are different for each "crime" and you simply cannot determine why people do something wrong or how to prevent it if you don't know what they're doing wrong. Ethical violations can be committed in many ways. Several specific violations are presented below. They are grouped by the type of violation in that some involve misuse of results or other data, while others are more related to intellectual property. Finally, there are "other" categories such as conflicts of interest, misconduct involving human or animal research, and misconduct during peer review.

- Let's call violations where results are involved crimes against science.
  - Falsifying and fabricating data
  - Deliberate omission of known data that doesn't agree with the hypothesis
  - Misrepresenting others' previous work done
  - Intentional negligence in the acknowledgment of previous work done
- Let's call violations where ownership or authorship of results are involved crimes against researchers.
  - Passing off another researcher's data as one's own
  - Publication of results without the consent of all the researchers
  - Failure to acknowledge all the researchers who performed the work
  - Repeated publication of too similar results or reviews
  - Breach of confidentiality
- "Other violations"
  - Conflict of interest
  - Violations where animals or humans are involved
    - Violations of human subjects research committee or animal research committee protocols and/or federal regulations
  - Violations during the review process
    - Peer reviewer violations

    What each of the violations is, why they happen, and how they are caught will now be discussed. When appropriate, the damages each cause will likewise be discussed.

CRIMES AGAINST SCIENCE

FALSIFYING AND FABRICATING OF DATA YOU'VE COLLECTED

What Is It?

The more colloquial way to describe fabrication of data is lying. It is one instance of scientific misconduct that is also, without question, a violation of the moral ethics of decent people. In this ethical violation, researchers quite literally make up data, claiming experiments were carried out when they were not or significantly altering the results they *do* obtain so that they fit the pre-experimental hypothesis or previous

CONSIDER THIS HYPOTHETICAL SCENARIO

FALSIFYING DATA

A researcher is searching for new planets using an infrared telescope, measuring how the heat signature of the stars decreases as the planets pass between our relative position and the stars'. It is a total failure, but they create a table of results that describes the effect being observed and write it up describing the identification of planets previously discovered, claiming it is proof the method works.

FABRICATING DATA

A researcher believes that planets can be discovered using an infrared telescope. The researcher claims that as the planets pass between Earth and the star, they absorb enough heat to be measured. The researcher calculates how much heat would be absorbed with Mercury and Venus as they pass between Earth and the Sun and then claim that using the infrared telescope, and is able to detect when the planets traverse the Sun between 25% and 75% with the maximum calculated absorption at 50%; meanwhile, no experiments were carried out.

studies. This ethical violation is among the most difficult to catch. This is because data fabrication can only be caught if another scientist attempts to repeat or otherwise use the research that was fabricated. This ethical violation is also perhaps the most damaging of them all—not just because it is also, undeniably, morally wrong, but because it also has the unfortunate consequence of leading other researchers down an incorrect and potentially **impossible** path trying to repeat and/or use the fabricated results. This then has many negative effects on other researchers' careers resulting in wasted time and research funds. In these days of intense pressure to publish and obtain grants in times of tight financial stress, losing time and/or money could result in grants being terminated or denied funding and/or in the destruction of a new faculty member's chances of being awarded tenure or promotion. Consequently, fabrication of data violations is often met with *severe* repercussions such as termination from a position or bans from applying for federal grants. Even with such serious threats hanging overhead, many people still succumb to the temptation for reasons that will be touched upon below.

Also falling into this category is alteration of results (especially spectroscopic results such as NMR) to make the data appear more like that which was expected or desired, or to make the product of a chemical reaction appear purer. Although some may argue that "It's no big deal, I only photo-shopped out the solvent," it is still scientific misconduct. It is most unfortunate that this is virtually impossible to catch. Many would argue that this is not as bad as quoting a crude yield of 96% while neglecting to report that the pure yield is 34%, and perhaps that's true though I cannot agree. These are, without question, *bona fide* fabrications of data. With the advent of more and more advanced computer programs, this type of data fabrication is becoming easier with each passing year. In fact, somewhat recently, the *Journal of Biological Chemistry* announced its adoption of the *Journal of Cell Biology's* policy because this has become a more prevalent issue that reviewers and readers must be more cognizant of than ever before.[1] The policy reads: "No specific feature within an image may be enhanced, obscured, moved, removed, or introduced. The grouping of images from different parts of the same gel, or from different gels, fields, or exposures must be made explicit by the arrangement of the figure (e.g. using dividing lines) and in the text of the figure legend. Adjustments of brightness, contrast, or color balance are acceptable if they are applied to the whole image and as long as they do not obscure or eliminate any information present in the original, non-linear adjustments (e.g. changes to gamma settings) must be disclosed in the figure legend." The Journal goes on to list the procedure that they feel will ensure the prevention/ detection of such misconduct. Their comments conclude with the following: "After due process involving the JBC editors, editorial staff and ASBMB Publications Committee, papers found to contain inappropriately manipulated images will be rejected or withdrawn and the matter referred to institutional officers."

Fabrication of data is significantly different from the data that is accused of being erroneous but not fabricated. I would argue that the latter of these cases are not an example of bad science or bad ethics but an example of scientific progress. Take as a brief example now the ancient belief that Earth was at the center of the universe. Based upon the evidence that was collected, this was indeed a logical conclusion for ancient humans to make. Modern evidence refutes this however, and we now know this to be incorrect. This, for certain, doesn't make ancient man's approach toward this conclusion unethical. It is also not bad science. They took the data they had and made what they believed to be the most logical conclusion. As science has progressed, we've become capable of not only proving that Earth is **not** at the center of the universe but also not even at the center of the galaxy or even our stellar neighborhood, the solar system. This is clearly a case of scientific progress and our ability to both collect and interpret data. Examples like this abound especially in medicine and are neither bad science nor scientific misconduct.

Why Does It Happen?

Why do people fabricate their data? Well, this one should be easy, even for someone who has nothing to do with any scientific field to answer. Many reasons, in fact, should come to mind instantaneously. For one, it is exceedingly difficult to catch, making it (potentially) very easy to get away with. Second, you'll very rarely, if ever, get a grant or publish a manuscript based on poor results, and without these (grants and publications), you're not going to keep your job very long and you will certainly have a difficult time earning tenure, a promotion, or a raise. Excuses for a pharmaceutical company to fabricate data are even clearer—many millions or even billions of dollars hang in the balance. The incentive to display good results is clear; your career and livelihood depend on it. The greed to be (or appear to be, more accurately) the best is a very tempting thing to some people.

[1] *Journal of Cell Biology*, **2004**, *166*, 11–15.

How Is It Caught?

This is one of the worst forms of ethical misconduct and one of the hardest to catch. It is simply impossible to catch this via the peer-review process (which will be discussed in detail later), as that would require a reviewer to check every claim.[2] With the enormous volume of work and increasing specialization of research, such a widespread effort is nothing short of unreasonable. Wholesale fabrication of data is impossible to catch before publication. Usually, instead, this violation is learned the hard way by an innocent researcher trying to use or further develop the fabricated results.

One way to potentially catch the digital alteration of results may be to take advantage of the improving computer technology that the perpetrators exploit and have raw data files (that have the appropriate time and date stamps) sent to reviewers. The reviewer could then use the appropriate software to recreate the figures and perform a check. This, however, would put enormous strain on the peer-review process, and, frankly, the point of peer review is to evaluate the science, not to detect fraud. This science is taken as being true, and this trust is essential to science. Furthermore, manuscripts that have a large volume of results have a correspondingly large volume of supporting data. Thus, it would be unreasonably lengthening the peer-review process. Such data could be made available in the supplemental information, however, so that all readers could perform this check it they desire.

DELIBERATE OMISSION OF KNOWN DATA THAT DOESN'T AGREE WITH YOUR HYPOTHESES

What Is It?

> **CONSIDER THIS HYPOTHETICAL SCENARIO**
>
> A researcher is studying the behavior and training/conditioning of dogs, monitoring their reaction to a particular stimulus. Of the 13 dogs in the study, 9 have the predicted reaction, while 4 do not. The researcher only reports the 9 showing the predicted reaction and makes no mention of the other 4.

This is very similar, but not identical, to the fabrication of data, especially since it may be better thought of as selective inclusion of data. It can be argued that this constitutes a falsification of data and this would not be incorrect. However, the action is different enough to warrant its own mention and category in my opinion. If any particular result(s) is/are left out of a publication, just because they do not agree with pre-experimental hypotheses, an ethical violation has certainly been committed. We are obliged as scientists to report all the data (without revision) that we obtain. (For sure, rounding to the appropriate number of significant figures is acceptable as

---

[2] Some journals, such as *Organic Synthesis*, do exactly this and are among the highest-profile journals as a result. However, detecting fabricated results is **not** the goal of *Organic Synthesis*.

performing appropriate statistical analyses.) Unfortunately, all too often there is a tendency toward leaving outlying data out and providing reasons why are touched upon below. Of the violations considered in this book, this may be the one for which coherent (but nevertheless still unethical) arguments can be made to defend it. That being said, this becomes a more egregious breach of code when statements in the publication suggest that failed or omitted experiments did, would, or even should work. Such a hypothetical scenario is discussed at the end of this chapter, in the context of comparing it to bad science.

There are tests, such as Student's t-test, Q test, and confidence limits, among others, that allow a researcher to determine whether a data point is insignificant. In these cases where a data point fails any of the aforementioned tests, it is entirely appropriate to omit the offending data. It would be most appropriate, however, to mention the data that has been omitted from the conclusion and the grounds on which it was omitted. This is because although it may indeed be an experimental/ equipment error, it may also represent an exception to your conclusion that may lead to a new direction of research. In cases where one data point from an experiment that was repeated 10 times is eliminated, these tests are particularly applicable. However, they are **not** applicable to disqualifying a single drug candidate when nine others are active, and the one outlier is significantly less active.

Why Does It Happen?

This is perhaps (and truly just perhaps) the only ethical violation that is defensible. It is most unfortunate that the peer-review process for the publication of manuscripts and funding for grants is sometimes not really kind to aberrant data. Even in cases where the aberration is unexplainable, it is almost less problematic to leave it out all together because often the reviewers red flag the data and use it as grounds to reject the paper or grant because the process isn't general enough or the new class of inhibitors has too many derivatives that aren't good enough. Unfortunately, this very practice of reviewers acts not only to facilitate dishonesty but also to hold science back. Therefore, many times, you must be dishonest (although perhaps not fully disclosing is more accurate) in this way to succeed. Widespread application of these sorts of peer reviews enables a culture where only the best results are published, rather than all of them. Full honesty is still the best policy, however. The more diligent reviewers, though not all, will require these results to be included and will not hold a negative result against the author. Full disclosure only shows the *limits* of a method, a demonstration that is easily as important as the method itself. In fact, it is far better for you to show the limits of your own research than for someone else to show the limits. At least in the former case, you can provide the reasons yourself for the "shortcomings" of your own work, retaining control of the narrative of your work.

How Is It Caught?

This is completely impossible to catch, as if someone doesn't report that they did something, there is no way to know that they did without some sort of legal subpoena of the work! The only way that this can be caught is if a diligent reviewer thinks of the very experiment or example that has been left out and demands it as additional work that must be done before the publication can proceed. This sort of demand does in fact occur. You may be wondering how this is not in conflict with the earlier statement that science is based on trust. Such a review and suggestion are indeed an evaluation of the science, a reviewer can suggest additional experiments that would further confirm or refute the authors' claim. It is part of the collegiality in science and a duty of peer reviewers.

MISREPRESENTING OTHERS' PREVIOUS WORKS

CONSIDER THIS HYPOTHETICAL SCENARIO

A research group publishes the results of biological testing of a new family of drugs aimed at treating Lyme disease. They report all of the compounds they've developed but only compare them to the previous compounds reported by others that are inferior.

What Is It?

Misrepresenting others' work done is indeed a different ethical violation than deliberate fabrication or falsification of data though it certainly can be construed as falsification of someone else's data. One example of this violation is to present your own (and aberrant) conclusions of someone else's work as theirs. Let us be clear about an important point: It is certainly **not** inappropriate for you to come to your own conclusion about another researcher's work. What is inappropriate, however, is if you present your conclusion as the conclusion of the original authors. You simply *must* be forthright in explaining that the conclusion you are presenting is your interpretation of the other authors' work. It would be most appropriate for you to also compare **your** conclusion to the one presented by the original authors, providing the appropriate context of your thoughts.

Why Does It Happen?

Misrepresenting previous work done by other researchers allows you to present your work as being superior to the other researchers' work. Because such a premium is placed on arriving at the best solution to a problem, it is clear that presenting your work as better than others' is going to be beneficial to you.

How Is It Caught?

Usually, this is difficult to catch. The only way in which this form of scientific misconduct can be caught is if the original author (or someone who is intimately familiar with their work) reads the new report and catches the "discrepancy." It can also be caught by a diligent researcher who reads both reports to best understand the study. Reviewers, remember, are experts in the field, and there is a reasonable likelihood they'd be familiar with the cited work.

Intentional Negligence in Acknowledgment of the Previous Works

What Is It?

> **CONSIDER THIS HYPOTHETICAL SCENARIO**
>
> A group of researchers is working on developing a new synthetic method. They find that it is superior to some established methods but inferior to others. In their manuscript, they compare their method to no other methods, at one point calling their work a first of its kind.

First and foremost, this is not to be confused with the *unintentional* negligence in the acknowledgment of the previous work done, which can be called poor science (something that will be touched upon later). It is not unethical to be ignorant. In this instance, a researcher is intentionally not mentioning the work that has already been done in the field. This does not necessarily always mean **no** previous work is mentioned. The same violation is occurring when an author only acknowledges the work inferior to his or her own but when he or she cherry-picks out of prior work to only compare his or her work to inferior work, it is (in my opinion) more misrepresenting prior work done than neglecting it. This is unethical because it then appears as if the author is the pioneer or a leader in the field. When this occurs in a grant application, it may result in misappropriation of funds by the funding agency. However, in a grant, it may cause the work to be viewed as having no precedent and, therefore, not a safe investment for the funding agency. This can therefore be a double-edged sword. It also provides no context for the reader. That is, the reader cannot compare the present to work to anything.

It can certainly be argued that it is not possible to track down every bit of work that has been done on a topic so that you can properly acknowledge all the work that was done prior to your contribution to the field. With the enormous volume of work being done today,[3] this is certainly a daunting if not an impossible task. Fortunately, however, there are electronic search engines that greatly facilitate these searches. Although most of us who use them regularly are certain that the databases that these search engines access are not complete, or infallible,[4] an honest researcher will utilize these tools to the best of his or her ability. It is their absolute responsibility to make at the very least a good faith effort to search all the relevant literature before proceeding with a publication or grant. Does this mean that you must read every published work in every language? Certainly not! Nor does it mean a physicist must read the table of contents of every biology journal to ensure that you are working ethically. However, if the very subject you are attempting to publish or study/research is the title of an article that is in a language that you are unable to read, you have to

---

[3] There are literally hundreds of journals for each of the sciences written in more than five different languages.

[4] If a typo is made in entering the paper or book into a database, you'll NEVER find it if **you** search for it properly!

better find a translator if you want to proceed ethically. For any other reason, you have to better find a translator to ensure that you're not trying to reinvent the wheel, something that is much more like bad science than bad ethics.

Why Does It Happen?

You may be wondering what someone must gain by doing this; a small divergence is necessary here to provide some background. There is a great prestige of having your work cited many times. This implies that your work is important or at least well read. A relative measure of this has even been developed; it is called the H-index. This H-index is a number that counts the number of references that have been cited that number of times. This is calculated in the following way: if a researcher has six publications that are cited six times each, he or she has an H-index of six. If, on the other hand, the researcher has ten publications that are each cited five times, he or she has an H-index of five. Taken to an extreme, if the original research is not properly acknowledged, the wrong people may get the credit for pioneering a field, and this can have disastrous repercussions, including eventual misappropriation of Nobel Prizes. In short, the stakes are **extremely** high to climb the mountain first. This violation essentially gives a false impression of superiority. The incentives are big to make it look like you were the first to climb that mountain, but that is often a fruitless effort as it is by far and away the easiest ethical violation to catch. If you leave out work that was previously done, especially if it is a field that is very competitive, it gives the appearance that the offenders' method may be the best or only one. As stated before, this prevents the work from being viewed in any context. This is, of course, downright misleading and *sort of* gives false credit. This putative false credit then has the effect of making the offender(s) look better than they really are and, therefore, potentially more likely to get a grant funded, or more likely to have the manuscript accepted for publication or cited more times. Furthermore, there is a difference between not mentioning **any** competitors and not mentioning only the best competitors while mentioning (i.e., citing) lesser competitors. It is absolutely a violation to only mention competitors that you are superior to, leaving out those who are superior to you. The only proper way to proceed is to show where your work fits into the whole picture.

How Is It Caught?

Ordinarily, this is an issue that is caught by the peer-review process and is an instance where peer review works well, as far as detecting scientific misconduct goes. However, I suppose that peer review cannot truly identify it as intentional; it nevertheless identifies missing prior work done. This is because the peer reviewers of grants and other publications are often, if not always, experts in the field, at least in principle and almost always in practice. They often are intimately familiar with the state of the art and who has made which contributions to the science. Consequently, a reviewer who knows of other work that ought to be referenced or included into the discussion, he or she is obligated to indicate so in their review and usually will indeed do so. If the editor takes the review to heart, he or she will inform the authors that the appropriate references must be added, unless the authors can defend why they shouldn't be.

CRIMES AGAINST OTHER RESEARCHERS

PASSING OFF ANOTHER RESEARCHER'S DATA AS ONE'S OWN

What Is It?

> ### CONSIDER THIS HYPOTHETICAL SCENARIO
>
> A researcher repeats an experiment found in a paper from 1892 and submits it
> for publication to a high-profile journal, without referencing the original paper.

Another gross ethical violation is the passing off someone else's work as your own.
This is like, but different enough from what was encountered in an earlier ethical
violation (*Intentional negligence in acknowledgment of previous work*) to warrant
it's own class. This falls into the category of plagiarism even if you repeat the work.
In truth, the complete appropriation of someone's work or the copying of your own
work is plagiarism. This does not have to be limited to publications or grants they
read/review or seminars they attend. Ideas can be stolen to, and to do so is as bad as
stealing the work accomplished. Very rarely does an author acknowledge in any way
a personal communication[5] that led to a breakthrough, though they always *should*.
The offender could also just as easily (and just as unfortunately) take ideas from a
grant they refereed and incorporate them into their own grant or research for their
own gain. In other cases, the work being stolen is hastily repeated and then the new
results written up, published, or incorporated into a grant. In the latter case, that the
work was repeated really does not make it OK. In fact, it perhaps makes it worse as
it demonstrates additional effort toward being dishonest. A prime example of this
was brought to light in 2006. This is the case of Armando Córdova of Stockholm
University in Sweden discussed in Part B of this book. Also, with the enormous number of chemical journals in print today, it is even possible to just resubmit someone
else's work *as is* to another journal for publication with different names on it even
after the original paper has been published. Millions of articles are published every
year, and it's truly impossible to check these many articles, despite the improvements
in computer-assisted searching. It is becoming more difficult to get away with, however, as more journals turn to plagiarism detection software, although the software is
not as proficient at catching reused figures or other digital images, for now.

The question that you may be tempted to ask at this point is: *Just what needs to
be referenced?* There is in fact an issue known as common knowledge. For example,
nowadays, we don't need to reference or cite in any way that all known life is carbon
based. Although that known life is carbon based is a fact that had to be discovered, it
is now common knowledge and no longer needs a reference. Similarly, nobody in his
or her right mind would cite that high cholesterol causes heart disease, even though
real research revealed that fact long ago. Likewise, nobody references Watson and

---

[5] For example, an idea that comes from a discussion in your office with a colleague or visiting seminar
speaker.

Crick's manuscript when describing the double helical structure of DNA anymore. A good rule of thumb is that if the general public (nonscientists) knows it, it can safely be considered as common knowledge and does not need a reference. A better rule of thumb is: *When in doubt, cite!* You are never wrong to be providing a citation for something that may not need it, although some journals prohibit some things (such as anything unpublished) from being cited.

Regarding presentations, seminars almost always contain varying amounts of as yet unpublished results. As mentioned before, occasionally, someone in the audience may think that something that is not yet published is publishable *as is*. They then may feel that if they could repeat the results, quickly, in their own lab, they may be able to scoop the original researcher and appear to get there first or even just publish the results directly from the seminar. This is most unfair as seminars are often given in confidence that the audience will not perpetrate such acts and is a horrendous breach of professional trust. Furthermore, department seminars are usually considered to be personal or private communications. Seminars that are at national meetings are an altogether different conversation. These are certainly a public disclosure, and they must be referenced just like a journal article should be. It is therefore completely inappropriate to attempt to report data from someone's seminar as data you've collected, even if you repeated the work. To do so, no matter the setting of the presentation is plagiarism. It is wrong, even if you cite the presentation.

You may be wondering what happens in the rare cases that researchers make a simultaneous discovery. This can lead to several different scenarios, some of which are ethical and the others unethical. In the ethical case first, we assume that the reviewing researcher is also pursuing the same goal and is preparing a manuscript of their own. This, of course, is a reasonable assumption since most peer reviewers are experts in a particular field. In some fields, there are very few experts. Hence, it is inevitable that a reviewer will also be submitting a publication reporting the same or, perhaps in some cases, different conclusions. From here, there are several different ways the reviewer can react. First, and most ethically, the reviewer would alert the editor to this conflict and recuse himself or herself, explaining the source of the conflict. The editor may then ultimately propose the papers be published in series with one another while being reviewed by someone else, since they are so closely related. In this way, neither researcher receives any significant benefit or detriment. However, it is important that the reviewer informs the editor *before* reading the paper. This is because the reviewer cannot "unread" the paper and may inadvertently or in a more sinister case deliberately use this data in his or her own research, which *would be* unethical. If the reviewer only realizes the close relation after reading it, he or she must inform the editor at once. Although it is too late from the point of view of reading and learning, his or her ability to objectively review the work is likely compromised, and he or she must contact the editor.

For sure, it is inevitable that a researcher will receive inspiration for a new project or critical new insights for an ongoing project during the review of a manuscript or grant submission. This is different from what was previously discussed, and these differences deserve mention. Here, it is the insinuation that the reviewer recognizes something not directly related that can still be applied to his or her own project. The case of reviewing a manuscript involves much more clearly defined lines and will therefore be considered first. The most appropriate way to proceed, if a reviewer of a manuscript identifies ways in which the work can be applied to his or her own work, is for the reviewer to immediately make this potential conflict of interest known to the editor. However, the editor chooses to proceed (whether the reviewer is replaced or allowed to continue), and the reviewer simply cannot "un-read" what he or she just read and learned. This immediately opens the potential conflict of "how long must I wait to act on what I read." The only ethical way to proceed is to wait until the manuscript has been published or otherwise been made available to the general public.[6] This removes any unfair advantage the reviewer would have potentially gained. The truly *most* ethical way to proceed would be to even have the self-discipline to not do any work using the results in the manuscript until then. However, many would likely consider that level of rigor overkill, and those who consider it such are wrong. Another potentially viable option *may* be to ask the editor for permission to contact the author and perhaps begin a collaboration. It is critical to note that this sort of contact **must** have the editor's approval, if it can be done at all, since reviews are ordinarily anonymous. After it is published, it is a fair game to contact the author. Even in an open review format, the editor should be consulted before a reviewer reaches out to an author in this, and we'll see later.

The situation is far more complicated when the reviewed work is a grant. Unlike nearly all manuscripts, which by and large are complete projects or at the very least complete mini projects, grants are *proposed* projects, even though they often build upon previous work by the same author(s). This not so subtle distinction makes it much easier to steal work from a grant since journal articles (manuscripts) usually serve to "stake claim" to a field, something grants simply do not do in any way. Furthermore, publications can be searched for, while unfunded grants can be less easily found, so no official record of the proposed idea survives in the commonly searched literature. Because of this, and because grant reviews are anonymous, there is no ethical way to use novel information found in a grant. Even and especially contacting the author is inappropriate. The only ethical way to proceed is to be patient. The situation can become even more complicated if, while reviewing the grant proposal, a reviewer envisions an equally novel, but superior quality idea based upon the same work. This is likely to be inevitable and, since *so* much work goes into authoring a grant proposal, if the reviewer then submitted his or her superior idea in the next round of proposals, it would be very difficult to convict him or her of scientific misconduct, and in fact this is likely a gray area. However, sometimes the court of public opinion is even more draconian than the real penalties, and this sort of behavior will likely be met with anger from persons in the scientific community.

A related concern that, in truth, applies to both manuscripts and grants is that the reviewer may come across a reference he or she never knew about. Like before, this cannot be "un-read." Reviewing grants and manuscripts is in fact more than just a service to one's field; it is also a way to learn. There would indeed be nothing unethical at all with using this new-found reference in the reviewer's own work.

In another related scenario, the reviewer of a manuscript may identify a way in which he or she can either improve the research or use the research to improve one or

---

[6] This happens upon acceptance for publication at most journals.

more of his or her own projects. Here, the potential for scientific misconduct is significantly higher. There are only a small number of ways that the reviewer can proceed ethically: first, he or she can be completely selfless and make recommendations to the author(s) based upon his or her ideas. Second, *after the paper has appeared in print*, the reviewer can prepare a publication that demonstrates these improvements. This, however, introduces a very gray area: "When is it ethical to start this sort of research project?" It is my opinion that this sort of research *should not* start until after the (original) paper has been accepted for publication. This is because at this point most journals make the publication available to the greater scientific community. Only at this point is the reviewer not getting an unfair advantage. Does anyone actually wait for this "grace period"? Probably not, though I'm not aware of anyone ever being convicted of this. Furthermore, it would be very difficult to get a scientific misconduct charge to stick in this sort of case.

One thing that would certainly be unethical would be for the reviewer to hold up the submitted publication while the "improvements" are made and then submit these improvements for publication. Such would, for sure, represent gross misuse of the peer-review system. This sort of violations is also nearly impossible to catch because peer review is anonymous. And, although the editor knows who the reviewer is, the reviewer can lie about the delay as they submit their paper to an alternate journal.

Another entirely unethical scenario can be envisioned if we make the assumption that the reviewer is not working in the specific area involving the publication. If the reviewer was to then immediately start a competing project based upon this publication, he or she is committing scientific misconduct. If such a reviewer intends to contribute to this field, he or she **must** wait until the publication has been made available to the greater scientific community. Even in this case, it is potentially very difficult to find the line between ethical and unethical. In all cases, one must be very careful when he or she reviews a paper to not incorporate someone else's data into his or her own research.

Why Does It Happen?

The why for this ethical violation ought to be obvious; a small discussion though is still worth it. Obviously, if a violator can pass off someone else's work as his or her own, especially if he or she competing toward the same goal, he or she can stand to gain quite a bit. If the work can be successfully stolen from the competitor, the perpetrator now looks superior.

How Is It Caught?

Catching this is similar to and in some cases identical to not acknowledging the previous work done—that is, the peer-review process can often, but not always catch it. Often, it may only be caught if one of the reviewers is familiar with the work in question or if a particularly diligent reviewer searches for the work he or she is reviewing. In cases where there is a high degree of poetic justice, the reviewer may be the original author of the work under threat of being stolen. Most often, it is only caught after the manuscript is published and the author of the original work or somebody who has read the original work happens upon the "new" manuscript.

PUBLICATION OF RESULTS WITHOUT CONSENT OF ALL THE AUTHORS

What Is It?

> **CONSIDER THIS HYPOTHETICAL SCENARIO**
>
> A member of the lab leaves for another position after a particularly heated dis-
> agreement. A manuscript they worked on is submitted to a journal with their
> name on it but they are not told about it until after it is accepted.

One often overlooked ethical violation is that all authors must approve the publication. Some journals have all authors sign off on a manuscript before continuing with
review or publication. However, such approvals can be and have been forged. When
it is overlooked, it is often an honest error and can be resolved peacefully. There are
cases however that can be imagined where one of the authors does not agree with one
of the chief conclusions the other authors are putting forth. In a case such as this, a
resolution **must** be reached before the publication can (ethically) proceed. There are
also cases where one of the authors may not even be aware of the paper, and this is
not unthinkable since sometimes parts of an author's work may get published years
after leaving the lab. This can also occur innocently or with more sinister motives.
In both cases, it is not the fault of the person who does not know about the publication. They have for sure committed no ethical foul in this situation. A problem can
result when the publication is brought to his or her attention after it is *in print*, only
then finding that they disagree with one of the fundamental conclusions a conclusion
that just for the sake of argument is in direct opposition to one they just came to in
their own (new) lab. In a case like this, it is easy to see why that person would be
little upset when they were not consulted. This situation could also cause significant
complications for the forgotten researcher when applying for a grant. In the National
Science Foundation (NSF) biosketch, principal investigators (PIs) are required to
list collaborators within a certain number of years. A situation can certainly occur,
although it should be rare, where a paper is published from work completed inside
this period. The granting agencies may then interpret this as an improperly completed biosketch and return the grant unreviewed and would constitute misconduct
on their part. Also, having the opportunity to approve or disapprove of the publication is important for a very practical reason. This reason is that all the authors,
by virtue of being authors, take responsibility for the work. Thus, if there is ever
anything about the work called into question, be it for bad science or bad ethics,
everyone involved carries that stain, at least to some degree.

Other reasons may compel a coauthor to object to the publication of a manuscript.
Chief among them is the author order and who is identified as the corresponding
author. Certain positions in the author list are hotly vied for. In many, but not all,
journals, the first author position is especially treasured. This position is usually
the one occupied by the researcher who contributed most to the work, the writing,
or both. Being identified as the corresponding author is also highly valued as this
makes a person the contributor whom people should contact if they have questions

or comments about the research. Effectively, it makes this researcher the mouthpiece for the project.

In gross cases, an author who may have nothing to do with the experiment is added to a paper to give it more "weight." For example, imagine adding a Nobel Laureate's name to your paper to increase its chances of getting published, and said Nobel Laureate having **never** even heard your name before. I've had private conversations with a reviewer for a journal where the reviewer expressed the opinion that manuscripts he or she felt were terribly written but from an eminent chemist's lab were still accepted with little or no alterations leaving the reviewer's feeling as if his or her review was ignored. The reason why it increases the chances of the paper getting published is that many reviewers and editors may be more hesitant to make negative comments on an article coauthored by such a preeminent scientist, and it can certainly be argued that this is an example of where peer review fails to operate as designed and may even be corrupted. In a case like this, it is the author(s) not the science that is evaluated. This should clearly represent an unacceptable practice.

Regarding author order, whatever the convention of the journal, this is an issue that should be decided before the publication is prepared in the first place to keep the issue from preventing or delaying publication or worse yet, retracting one. While in most cases, direct conversation and openness is the best approach, it is not always straightforward and may require equally contributing coworkers "taking turns" being the first author in cases where multiple papers are *legitimately* published by the same team on the same or related projects. One of the ways this has been resolved is to add a footnote that indicates certain authors contributed equally to the work.

Why Does It Happen?

Why this happens depends on whether it is the innocent variety or the sinister variety. Considering the innocent variety first, the fact of the matter is that some people don't fully appreciate that it is unethical to list someone as an author for a piece of work and not run it by them for approval. For sure, most PIs want the input of all the workers involved, but some people consider this an option and not an obligation.

A PI may want to keep someone in the dark about a publication because the PI thinks he or she may object to something in the paper. You may be tempted to say that the PI should be commended for adding such a worker as an author to begin with, but I can't agree. You can't make one ethical violation OK by going out of your way to not commit a different one (and not crediting the author is a violation that is discussed next). Life and science just do not work like that. Scientific misconduct is not like acquiring credits that allow you transgressions!

Finally, take, for example, adding a famous researcher in the field onto the author list. This, by simple virtue of association, makes the paper a stronger paper. Whether this is "the way it should be" or not is irrelevant in this discussion—it's true. The likelihood of the paper being rejected or harshly criticized is immediately and drastically reduced, just by the simple inclusion of that famous researcher's name. Thus, the incentive is to have you work less criticized.

To demonstrate this unfortunate tendency, consider the two cases below:

One instance where having an expert whose work is perceived as being beyond reproach may be the fraudulent claims of the discovery of element 118 by the

Lawrence Berkeley National Lab as reported in 2002.[7] In 1999, Victor Ninov, a then researcher at the Lawrence Berkeley National Laboratory, was found to have fabricated results that suggested the formation of element 118. Follow-up experiments at Berkeley and other labs in Japan and Germany failed to confirm the results, and an investigation revealed Ninov's misconduct. The paper was retracted in 2001, and after an investigation, Ninov was fired. Around the time Ninov was fired from the lab at Berkeley, previous work he performed at the Institute for Heavy Ion Research (GSI), Darmstadt, Germany, was scrutinized. Although GSI validated the claims to elements 110 and 112 that Ninov worked on, they discovered some of the data was fabricated and did not exist in their records. When an expert claims a discovery, particularly an expert at an upper echelon institution such as Berkeley, few, if any, question the results, and that is sometimes taken advantage of.

Also, when fabricated results confirm our prediction(s), the tendency to believe the results is that much stronger. This has been pondered[8] as a contributing factor in the case of J. Hendrick Schön of Bell Labs, who was found to have fabricated results in the area of superconductivity and molecular electronics. If previous reputations did in fact contribute to either of those works being published to begin with (and to be fair, they may not have), the point is clearly illustrated; famous researchers are given more benefit of the doubt and their publication road is thereby easier.

With the high turnover of researchers (especially outstanding undergrads), it is not uncommon for researchers to lose touch. This could make contacting potential coauthors problematic, if not impossible. Social and professional networks like Facebook and LinkedIn should be harnessed to alleviate this, and if multiple and varied attempts to reach a coauthor fail, the editor should be asked for advice.

How Is It Caught?

Ideally, this is caught easily and handled peacefully; the uninformed author finds out, forgives, and publication proceeds with the author's approval. In other cases, the now informed author may object to and successfully block publication or even have a manuscript with sound science retracted. If the publication is blocked, the issues causing the objection must be resolved before publication can move forward simply removing the author, even if their work is removed too is **not** an acceptable option.

FAILURE TO ACKNOWLEDGE ALL THE RESEARCHERS WHO PERFORMED THE WORK

What Is It?

CONSIDER THIS HYPOTHETICAL SCENARIO

A graduate student leaves upon graduation for the next stage in his or her career. The students continuing the work started by the student who graduated complete it and publish the work without the first student as a coauthor.

---

[7] Jacoby, M. *Chemical and Engineering News*, 4 November **2002**, 31–33.
[8] From *J. Chem. Ed.*, **2002**, *79*, 1391.

This is the deliberate omission of deserving authors from a manuscript and is quite different from the previously discussed violation; in fact, it is more similar to plagiarism as you are using their work as your own. This is one of the ethical violations that are subject to a great deal of gray area, perhaps even the largest amount of gray area although the hypothetical case here is pretty blatant. For sure, some journals and patents have well-articulated rules governing authorship from the point of view who is an author and the author order. Some journals go so far as to have the authors indicate each author's contribution. However, these rules are not necessarily interpreted consistently. Obviously, enormous fights could be waged on whether someone has added enough to a project (intellectually, or lab work) to warrant being a coauthor on a paper or a coinventor on a patent and with the high monetary stakes involved with especially patents, which is an enormously important and sensitive issue. It is in the best interest of the PI to be as fair as possible to all the (potential) authors in these cases. If he or she gains a reputation for not granting deserving workers coauthorship, they will reduce the likelihood of quality workers wanting to work for him or her, and an ethical violation would certainly have been committed in not giving appropriate credit. This reputation has this effect because potential workers will be concerned that their own significant contributions to a study may go unacknowledged. If a worker sees evidence that his or her hard work may not earn him or her any credit, he or she will inevitably search for a situation that will ensure being given the credit he or she deserve for the work he or she does. It is also important to not be too generous as adding undeserving authors has the effect of diminishing each author's *apparent* contribution.[9] Including undeserving authors has this effect because it gives the appearance to the general community that it was an enormous team effort with all the authors contributing to at least one essential component of the study. This logically reduces the perception of every individual's contribution. An example could be a PI including as a coauthor of a particular work technical staff who upkeeps laboratory instrumentation or provides critical insights into the interpretation of the results. While the researcher who operates the instrument or assists in the interpretation of results should absolutely be included as a coauthor, the researcher who merely performs routine maintenance should not. An exception to this rule of instrument operation is if you do a favor for a friend and run a sample on an instrument or a simple purification for them **once**. In this sort of case, including you as a coauthor is probably not appropriate, though including your assistance in the acknowledgments probably is.[10] There is an exception, however. If this is an instrument that you designed or that your friend does not know how to operate, then you are indeed deserving of coauthorship, though some may say this falls into a gray area.

One final case is when a potential author is intentionally left off a publication due to a personal falling-out, a very major violation has occurred. Once again, publishers, in some cases specific journals, professional organizations, and patent offices, have established rules regarding authorship, and the reader is encouraged to consult

---

[9] An interesting situation to consider is one where a PI requires all workers to relinquish all rights to patents. Is this ethical?

[10] At the very least, dinner is owed.

his or her professional organization's or journals' rules to clarify any confusion that arise during their career.

Why Does It Happen?

When deliberately committed, reduction of the author list makes it look like those remaining authors did considerably more work than they really did, and it harms the omitted authors as it denies them credit for the work they've done. For patents, reducing the author list leaves a larger portion of potential royalties left to the remaining author(s). While the PIs ought to keep careful track of who does what work on each project, they are **human**. They forget things, they lose track of things, and they (often) have several other projects with large groups of people. In these cases, it is often caught and handled with general peacefulness as there is a high likelihood that the omitted author is still working in the same lab or in some way still holds a professional relationship with the PI. When it is not caught before publication of the manuscript, a corrigendum can be submitted that adds the author(s) or a paper may have to be retracted.

However, there can be cases where the motives are more sinister. Take, for example, if there is a falling-out, and the PI wants to get revenge on one of the workers, he or she may be tempted to leave them off the author list of a publication. In this sort of case, the incentive is more like revenge or perhaps an attempt to avoid the conflict of another collaborative effort with a researcher that there is a poor working relationship. This is a clear and unfortunate ethical violation. Also, sometimes, a post-doc, a graduate student, or a collaborator may be bitter that the advisor or co-PI doesn't want to publish the results yet. If any of the other researchers attempts to publish the work, simply leaving off the coworkers who don't want to publish yet, very gross scientific misconduct has occurred. This process is somewhat tied to the violation of not obtaining permission from all the authors. It is different here in that the objecting authors are left out altogether though their work is not. That is, they are not added but kept in the dark about the publication or their objections are blatantly ignored. Even if the work is repeated and the new worker included as an author, the original author **must** still be included since processes they developed or otherwise significantly contributed to were likely used.

How Is It Caught?

This is also usually caught relatively easily. When an author sees an article that he or she thinks it should be listed on, he or she will make those feelings known. This is greatly facilitated by the fact that sometimes, all those involved are still working together in the same laboratory. In cases where the omitted author is no longer associated with the lab or anyone in it, however, it is much less likely that this will be caught before it is too late.

Authorship and Intellectual Property

The two previously mentioned infractions can have other important consequences that are logical to point out now. One reason it is so important to, for example, have the consent of all authors and to make sure all deserving authors are accounted for relates to intellectual property. The former can then be related to or perhaps more accurately, inadvertently give rise to a Conflict of Interest violation.

One way in which procuring the consent of all the authors relates to intellectual property is patents, which, for some products, may net the university or other institution millions of dollars. In such cases, the authors of the patent, *the people who actually made the discovery and/or invention*, almost always share a piece of that windfall. Sometimes, these royalties may precipitate a conflict of interest that goes beyond the one previously mentioned with NSF bio-sketches. At the very least, such an arrangement would have to be declared at most academic institutions. If a researcher is placed onto a patent and does not know about it, the appearance of being negligent in this declaration will potentially cost that researcher his or her job and may result in fines or even time in prison.

Leaving deserving authors off, a patent is obviously something, considering the immediately preceding discussion that can be immensely damaging to not just one's career but livelihood as well. It is therefore one of the highest obligations of a PI to keep very accurate and up-to-date records of exactly which coworkers and/or students performed every part of every study.

REPEATED PUBLICATION OF TOO SIMILAR RESULTS

What Is It?

> **CONSIDER THIS HYPOTHETICAL SCENARIO**
>
> A researcher submits a manuscript to one journal as an invited review, and then submits the same manuscript with minimal updates to another journal within 1 year. Both are accepted and appear in print a few months apart.

Often, researchers try to boost their publication records by publishing multiple times on the same work or too closely related work. This is not unlike plagiarism and in fact is a form of plagiarism called "self-plagiarism." This has the result of giving the impression that far more work is being done than is. One way in which it occurs is when the same project with no or minimal new results is submitted to multiple journals. It can also be argued that when a few (three to five, for the sake of argument) published papers are combined into one large paper, with little to no added insights, the same violation is at work; instead of, an argument I would agree with. There are also scenarios that can be conjured up that would make such a perspective paper appropriate, however. There are also instances, where breaking up work and/or publishing follow-up studies are appropriate. Consulting colleagues and editors is a good way to make sure you are making the right decision regarding your work.

For example, let's consider a hypothetical situation where a research group has investigated identifying different types of stars using a new type of telescope. It would be far more appropriate to include all the results in one paper rather than to fragment the research into separate publications detailing how the telescope was used to identify each type of star. That being said, if there is a type of star this new telescope can identify that others cannot, or maybe even very few others can, this would certainly be appropriate to publish alone, while incorporating the remainder of the work into a second paper. This is appropriate because it (1) reports all the work and (2) places appropriate emphasis on something that is novel and may otherwise be lost in the details of the other results. The authors would be incorrect, however, to not have whichever manuscript was published second cite the one published first. The same can be said about a new synthetic method that leads to a family of compounds that are subsequently found to have a potent anticancer activity. Publishing these as separate papers, one that focuses on the method and the other on the biological activity, allows the appropriate focus to be placed on both breakthroughs. If it were not a new synthetic method that led to the new compounds, this would likely be unethical.

Reviews can be a sticky situation with regard to multiple publications. For sure, annual reviews on the state of the art in a field are not inappropriate, provided of course enough new work is done over the course of 1 year or if the annual review focuses on the last year only. This however can be abused if an author writes multiple related reviews for different journals, even as many as 2 years apart. Even with how fast science moves today, it does not move quickly enough to warrant multiple reviews on the same topic each year in most cases. To be clear, it would not be unethical to write reviews for each of the different HIV inhibitors in 1 year. What *would be* unethical however would be if this same author wrote a large summary review of HIV chemotherapy, using large amounts of the other, smaller reviews with little or no changes in this hypothetical larger review. First, it is self-plagiarism. In using the smaller reviews in the larger one, the larger one does not contribute anything novel to the scientific community. Even if the previous smaller review is cited in the larger one, taking large amounts of work from a publication with little to no changes (even your own work) is not ethical and constitutes scientific misconduct.

Why Does It Happen?

Why do people repeatedly publish similar results? As mentioned earlier, such behavior increases a researcher's publication record. At most universities, especially major R-01 institutions, one of the most influential factors that influence the promotion tenure decisions is the publication record of the candidate. The term that is often used to describe the premium placed on publications is "publish or perish." With such stakes on the line, it is no small wonder that (especially young investigators) feel enormous pressure to publish and at a prolific rate. Not surprisingly, this pressure eventually leads some to cut corners and begin a march toward the line of bad science vs. bad ethics, a line that some inevitably cross. Which will be weighed more heavily, the number of publications or the quality of publication, is something that may vary from school to school and maybe even from committee to committee within a school.

What was said about promotion and tenure can also be said for getting grants. No granting agency will award a researcher a grant in the total absence of evidence arguing in favor of the potential for success. Some reviewers will even consider the author's perceived level of expertise in the field. Usually, this expertise is measured by the number or quality (or both) of *recent* publications by the author in the field. Since without funding, little, if any research can be done, the extremely high emphasis on one's publication record is obvious to all in the profession. Thus, this ethical violation is often perpetrated to either improve one's job status or even get a job.

How Is It Caught?

This is perhaps the easiest to catch. If when searching a topic there are multiple "versions" of the research already published, the peer-review process and a diligent reviewer should reject a paper on these grounds and a good editor will adhere to that rejection. Also, in most cases where a researcher is perpetrating this violation, he or she nearly always cites his or her own previous work.[11] Once again, a diligent reviewer will catch this simply by doing his or her duty as a reviewer. In this case, citing the work does not absolve one of the self-plagiarism. Another case in which this can be caught fairly easily by the peer-review process is if an author submits the same manuscript with little or no changes to multiple journals. Recall from earlier that experts in the field are the reviewers! It is common that a researcher is a reviewer for multiple journals or grants. In this sort of case (where one reviewer reviews the same paper more than once for two different journals), the violation is clearly identified and there is no defense.

When applying for a research grant from the NSF, the authors must list all other grants that are either active or under review. It should be noted here, however, that grants unlike research reports can be submitted to multiple places with minimal alterations to fit the specific scope of any one organization. However, some organizations, for example, the NSF, require authors to disclose all active and submitted grants. Although this is not necessarily done as a check on multiple submissions (it is actually done at least in part to see if the author has the *time* to do the proposed work), it can function in this capacity as well.

Breach of Confidentiality

What Is It?

> **CONSIDER THIS HYPOTHETICAL SCENARIO**
>
> A reviewer, upon reviewing a manuscript related to a colleague's work, shares the manuscript with his or her colleague.

While some people look at this as "healthy competition" from the point of view of trying to steal something like trade secrets at least, they couldn't be more wrong. Many major companies have their employees sign agreements that prevent them from going to the company's competitors and allowing the companies to benefit from research or "insider" knowledge. This is, of course, very similar to stealing someone's work (Don't misunderstand… you can switch jobs and take your experience with you, you just can't take your research or other proprietary information with you!!). These confidentiality agreements are signed for a reason. If we didn't hold to these ethical standards, *company A* could pay someone to go work for *company B*

---

[11] Everyone cites his or her own previous work. It is necessary and in fact more unethical to **not** do so as neglecting to do so represents a failure to acknowledge previous work done.

for a few years and then come back with all its most important information. This is **not** "healthy competition," but rather a breach of an honor code, an ethical violation, and is the scientific equivalent to insider trading. If no agreements have been signed, the argument can be made that no violation has occurred. Legally speaking, this is probably indeed correct. However, the reality of the matter is that this would still represent scientific misconduct as the professional code goes above and beyond the legal code in this and other examples.

Another brand of conflict of interest or breach of confidentiality may be a reviewer not keeping information for himself or herself but passing it on to a friend or colleague who is a competitor with the work under review. Still another but less damaging to others is sharing something you are reviewing without consent.

Breaches of confidentiality are also discretely different from gamesmanship. To start out simply, let's use a sports-related case of gamesmanship. Former football player, Peyton Manning, quarterback for the Indianapolis Colts and later the Denver Broncos, was renowned for his ability to call plays at the line of scrimmage. Some of his opponents (especially the defenses of the Baltimore Ravens and the Tennessee Titans) have played against Manning so long that they've become familiar with the audible calls he used. Manning has countered by not only changing what some of the audibles mean but also employing fake or "dummy audibles" to keep the defense on a different page from him. Another example from the sports world involves players who are traded and use their knowledge of the former team's signs or a former teammate's weaknesses to gain an advantage for his or her new team. Nobody would consider either of these issues a violation of some ethical code. Likewise, it is not a violation of any kind for one university to investigate how another university does something like advising and then clearly articulate to perspective students why it employs a superior tactic to what the competing universities offer their students. Coming back to the world of science, if a former *Pfizer* employee informed his new place of employment, *Johnson & Johnson* that *Pfizer* had plans to hire 100 new research and development chemists working in the heart medication unit, it would certainly not be a breach of ethical codes for *Johnson & Johnson* to go out of its way to make sure its best scientists in that unit were not wooed away by *Pfizer*. Although it may be an excessively aggressive reaction, it likewise would not be a breach of code for *Johnson & Johnson* to also hire new chemists for that unit, trying to keep the best and brightest away from *Pfizer*. One can argue that this may represent a gray area and I would have a difficult time arguing against such a point. On the other hand, sharing information about specific research endeavors is not allowed; this is not gray.

Why Does It Happen?

Breaches of confidentiality happen for slightly different reasons than the other cases. Usually, breaches of confidentiality occur because someone has multiple loyalties and tries to improve his or her standing in one of those loyalties. Often, there is some sort of financial attachment, which is a clear and easily understood incentive. Other times, different incentives may be presented including material goods, advanced placement or positions, or a promise of favors in the future.

How Is It Caught?

Breaches of confidentiality are often caught when a company or other entity notices that its competitor suddenly has something very similar to something they are investigating/producing. Of course, a researcher can change his or her job, leaving one company (e.g., Pfizer) to go work for a competitor (e.g., *GlaxoSmithKline* [GSK]). What would immediately set off red flags, however, would be if a specific compound or even a family of compounds being developed at Pfizer (or even investigated but subsequently rejected by Pfizer) are suddenly under development at GSK after the job change. In such cases, Pfizer would certainly have grounds to make formal complaint. When this occurs, if there was indeed a breach of confidentiality and not just a common good idea, major litigation may follow. This can especially occur if the person committing the breach signed some sort of contract forbidding the transfer of intellectual property. Communications such as email may be subpoenaed and examined to prove a breach of confidentiality has happened.

OTHER TYPES OF VIOLATIONS

CONFLICT OF INTEREST ISSUES

What Is It?

> ### CONSIDER THIS HYPOTHETICAL SCENARIO
>
> An environmental scientist at a major university is called upon to be an expert witness for the defense in a case where a fracking company is being sued for polluting a nearby stream. The scientist neglects to disclose major stock holdings in the company being sued.

A conflict of interest can occur when a researcher or lab coordinator has some sort of vested (personal or financial) interest in a study, such that it may influence any number of things including, but not limited to, how long a failing project is continued; whether results are patented or published in peer-reviewed journals; or in egregious cases, even who is permitted to work on a project. One can also envision a conflict of interest exists when a professor writes a book that is required for a class he or she teaches. Such a textbook does not, immediately represent a conflict of interest. Clearly, a textbook that is written for a course by the instructor would be written in the same style and order that the material is taught by this individual. Therefore, rather than generating a true conflict of interest, an instructor-authored text can improve teaching and learning. However, it is certainly worth pointing out that a perfectly coherent argument can be put forth that contests, without using ethics and only using sound pedagogy, that this would have the opposite effect on teaching and learning. Such a discussion is not appropriate here.

Conflicts of interest are perhaps best illustrated by some additional examples with discussion. Imagine that a pharmaceutical company and an academic lab are collaborating on a study, with the pharmaceutical company providing financial support for the study, the fourth-year graduate student's salary, and a royalty-based stipend to the professor if the product ever goes to market. Assume that during the study, the student discovers a new synthetic method and that this synthetic method would have impacts that reverberate through the entire field of synthetic organic chemistry (ignore the hyperbole for the sake of this case). To develop this method (which would benefit the graduate student's professional career greatly) would mean that he or she would have to divert part of his or her effort away from the work with the pharmaceutical company. The research advisor *may* order the graduate student not to do this for fear of losing the additional royalty-based stipend. Let's be clear—if the appropriate institution paperwork is filed, the stipend is not unethical, the clouding of the judgment it causes (if it happens) is what is unethical.

Conflicts of interest can also be present in governmental institutions such as the NSF or National Institutes of Health (NIH). If a program director for the NSF has a sibling applying for a grant in that program, a definite conflict of interest may exist. Similarly, if an employee of the NIH has millions of dollars invested in a pharmaceutical company that is under investigation, it may influence his or her ability to do his or her job well. The same can be said of someone at the FDA.

Earlier, the topic of an instructor-authored textbook being used for a course was broached. For certain, the use of such a book is not unethical at most, if not all, institutions, with the requisite paperwork being filed. If, on the other hand, a student is required to purchase such a book and the instructor then does not teach out of the book or assign any coursework from the book, a violation has almost certainly occurred. In this hypothetical case, the instructor appears to be only mandating the purchase to add to his or her pocket. There are mechanisms in place to protect the students from this sort of conflict of interest. One example is the University of Connecticut (UCONN). If a faculty person authors a book, it must be approved for use in a class by a departmental committee to retain the royalties. In the absence of such approval, the royalties (at least a percentage of them, correlating to the number of UCONN students purchasing the book) must be donated to the University foundation or some other charity.[12]

In another case, let's assume that a research advisor has his or her own start-up company to market a material made in his or her lab. In this sort of scenario, it may be easy to fall into the trap of diverting more students' efforts toward a project that may give the professor great financial gains at the expense of the students' careers. One example of when this would have the most impact would be in the decision to author a patent over a journal publication. It is important to understand that once a body of work is published, it can no longer be patented. This is important because at this point (the point of no patent being possible), it is far more difficult to reap financial benefits from the work. Although in some cases, this is not an issue, in cases where a company or individual investor (even the PI of the study) is supporting the research, maintaining the potential for a return on this investment is of critical importance, and it is certainly not unethical. In these cases, a patent disclosure would nearly always be submitted prior to a traditional, journal-based manuscript.

---

[12] Personal communication, University of Connecticut Office of the Vice President for Research.

Where this becomes a problem is in that patents typically take much longer to prepare since a legal team or assistant must be involved to confirm the report is novel and file the requisite paperwork with federal and world patent offices. In some cases, after the work is patented, the PI and the researchers involved may sign over the rights to the work, causing all potential for publication of the results to disappear or at least be greatly delayed.

Conflicts of interest can also occur in a way that helps the author of a paper or grant. On all NSF grant applications, applicants must submit a list of collaborators and affiliations along with every grant submitted. The purpose of this is to ensure not that the author/PI is treated fairly, but to ensure that he or she is not given undue favorable treatment. It is for this reason that such relationships absolutely *must* be disclosed. For example, a grant proposal would not be reviewed by one of the authors' PhD advisors or former students.

In a scenario that doesn't necessarily involve science, many universities offer as a benefit reduced or even remitted tuition for family members of employees, including their spouses. This inevitably causes the scenario to arise where an instructor will have in his or her classroom or research group their child or spouse. At small universities, in a small town where many people don't leave town during their lives, other family members (i.e., nieces, nephews, etc.) may end up in that class too. This can be resolved a few different ways including mandating or recommending (or anything in between) that the student in question takes the same course with a colleague. In cases where this is not an option, alternate provisions for the evaluation of the student work (grading) can and probably should be made. Whatever the case, it is of the utmost importance that all evaluation guidelines are clearly established and documented. The application of these guidelines should also be clearly demonstrated. It would be very wise to take advantage of modern technology and create a digital archive of all work when this potential conflict of interest is present. It is important to note that in these cases, extensive and accurate documentation is critical to prevent the appearance of unethical behavior when it truly is not occurring and to effectively defend one's self if accused.

Why Does It Happen?

The easiest explanation for why conflicts of interest continue to occur is that to prevent them, one must go against the natural and understandable tendency to put a highest premium on one's own interests. For sure, it is no easy task to avoid this act of scientific misconduct. In fact, it may be the most difficult to avoid as most of us do not even know when we are committing it. As an example, it would be very difficult for just about anybody to fairly evaluate a grant or publication from a competitor, who just for the sake of argument is working toward the same chemical target from a different pathway. It would indeed be hard for most of us to fairly evaluate such a proposal. In a case like this, the reviewing scientist must inform the editor that he or she has a conflict of interest and therefore must decline further involvement. This is the only ethical response, no matter how tempting it may be to hold up or reject the publication or grant to suit his or her own interests. For certain, the editor of the journal or the program director of the grant could do a better job in picking reviewers, but this is not without its pitfalls. The primary pitfall of this approach is that the peer-review process is critically dependent on experts in the field reviewing the work under the current system. A publication or grant application will inevitably land in the hands of someone competing with the author. Furthermore, it is impossible for the editor to always know the specifics of what all the reviewers are working toward.

It should be noted that most journals do allow authors to request publications **not** go to certain reviewers. This allows at least some measure of protection against this issue. It is indeed exceedingly difficult for even the best of us to turn down an opportunity to gain an edge on our competitors or some (and in particular cases significantly more than some) extra money. When an advantage can be gained that, for example, prevents a competitor from receiving funding or delays/prevents publication of their work, some people may succumb to the temptation. Money is again at the root of the issue when things like instructor-authored texts are used in a class, financially benefiting the instructor. Familial, romantic, or sexual relationships with students or subordinates also all have clear and tangible benefits that cause the integrity of the scholarly or educational systems to break down all the while benefiting the offender.

How Is It Caught?

Catching a conflict of interest is much harder than it seems at a first glance. It is not as simple as abandoning a project where a *potential* conflict of interest may exist at the first signs of strife. Strife is *rampant* in science, and virtually every scientific discovery of any worth throughout all of history faced strife at some point (Just think of those who insisted the world was round against the voices of those who *knew* they would fall off the planet if they sailed too far or how many tries it took Thomas Edison to get the first light bulb to work.). It can very easily be argued that continuing any project that faces strife is simply the good, scientific thing to do. However, sometimes a researcher may be needed reminding or he or she may need to be pointed out by a trusted colleague that his or her professional opinion is being overwhelmed by his or her personal desires or goals. In these cases, most rational people will take these sorts of opinions of their colleagues to heart. The other scenarios mentioned above are more difficult to catch for various reasons, and in fact, the easiest way to catch it is accidental, or if someone makes an accusation and an investigation ensues. Also, conflicts of interest that influence someone's grant or manuscript review are usually tough to catch. The only way to catch them is for a program director (for a grant) or an editor (for a journal) to notice that a reviewer rejected something and is now trying to publish something that competes with that which he or she rejected. Sometimes, the best way to catch something is to put in place mechanisms that prevent them like the book authorization at UCONN.

VIOLATIONS DURING PEER REVIEW

Peer review is an important step in the scientific process. As such, an entire chapter is devoted to it later in this book. Know now, however, that there are rules and regulations that govern how it is done. There are also soft spots in peer review that allow for misconduct. Unlike all the types of misconduct previously covered, these soft spots entail misconduct by the reviewer, rather than the researcher—by the person(s) reviewing the work rather than those who did or are reporting it. For all these reasons, coverage of this will come later, rather than here. The primary audience of this book is less likely to be a reviewer in the immediate future, including the discussion here that would distract from the primary mission of the book.

CITATIONS AND PROFESSIONAL SOCIETIES

The different fields and journals in all manners of scholarly work, from the humanities, to the arts, to the physical, natural, and social sciences, and to engineering, all have their own style for citations. Guidelines for citations are available from an individual journal's "guidelines for authors" and/or from the professional society or publisher that owns the journal. Professional societies also publish their code of ethics that all of their members are expected to abide by.

Citations in Manuscripts

Regarding the format and style of references (citations), there are almost (and this is a little hyperbole) as many styles as there are journals. Some professional societies, such as the American Chemical Society, publish a summary of such and other guidelines for publication in the journals they manage. Such guides are very helpful to the preparation of manuscripts in a format that will be accepted by the editors. It would be wise, therefore, to review such guidelines before even beginning to write a paper.

There are also some reference manager programs commercially available. These programs work with Microsoft Word or some other word processor and enter references into a manuscript using the format chosen by the user. The user can create his or her own or use one of the many presets that follow the guidelines of many different journals across many diverse fields.

Some journals, such as the ones from the American Geophysical Union, as of this writing, stipulate that "Every reference must be available publicly online or in print before a paper can be accepted."[13] This means that in these journals, unpublished work or personal communications would not be allowed as references. The website even explicitly disqualifies "in press" work from being referenced. On the other hand, the American Chemical Society (ACS) provides guidelines for citing unpublished work and personal communication.

Professional Societies

Like with the peer review, most professional societies have their own articulated code of ethics for their members. There are too many individual societies to list them here, especially if you consider those abroad in all the developed nations of the world. Thus, no attempt will be made to list the societies, much less their ethical codes here. The reader is encouraged in the strongest possible sense to investigate their professional society's code. If you are unable to find it, ask a mentor. Most of them are likely very similar, but there will inevitably be some considerations more relevant in some fields than others.

---

[13] https://publications.agu.org/author-resource-center/text-requirements/, last checked 6/23/18.

LIVING RESEARCH SUBJECTS

Also, discussion of humans and animals in research will be given their own coverage since they more logically are stand-alone topics than any of the other forms of misconduct, if for any other reason, there are more rules for them than any other.

Related topics that do not necessarily represent scientific misconduct are the differences between bad ethics and bad science, proving previous results incorrect (scientific progress) and the whistleblower's dilemma. In each of the cases, the consequences can be as bad or worse than an ethical violation, though, in the latter, there are legal safeguards intended to avoid this.

BAD ETHICS VS. BAD SCIENCE

A very *very* difficult task in many (though not all) cases is discerning between bad *ethics* and bad *science*. Before wrapping up this chapter, this issue ought to be touched upon. For this, a made-up pair of scenarios is perhaps most instructive.

Scenario 1:

Frankie is an assistant professor at a major university. He is in his fourth year, and he is beginning to panic about *tenure*. His lab has developed a method of adding a selenium nucleophile to alkyl halides, resulting in the displacement of the halide by the selenium nucleophile. All the principles of organic chemistry suggest that if this reaction works well for alkyl halides, it ought to work even better for benzylic or allylic halides. However, when the work is done in his lab, the reaction fails. Frankie omits this data and decides in the paper to say, *despite this failure*: "When the high efficiency of this reaction is taken into consideration along with the fact that benzylic and allylic halides are often even more efficient substrates in similar reactions, we have every reason to believe that the same will hold true in these cases as well and an investigation into this is under way."

Let us take into consideration if something else happened instead.

Scenario 2:

Vladomir is a third-year graduate student at a major university. To have any shot at the ACS fellowship he plans to apply for, he needs at least one more publication to be in preparation or in print. He develops a method for adding selenium nucleophiles to alkyl halides. Overrun with excitement, Vlad convinces his advisor to publish, and they publish the work they have with the intent of exploring the benzylic and allylic halides next. They make the statement: "When the high efficiency of this reaction is taken into consideration along with the fact that benzylic and allylic halides are often even more efficient substrates in similar reactions, we have every reason to believe that the same will hold true in these cases, as well and such studies will begin shortly."

In Scenario 1, it should be clear after the earlier discussion that an ethical violation has occurred. Information was known by Frankie that directly conflicted with his "conclusion," and he intentionally withheld it from his publication. This falls into the category of deliberate omission of known data that doesn't agree and is an example of gross ethical misconduct. This is the case, even though the claim that "investigation into this is under way" was made. For certain, this can be a defense of why no violation has occurred. Frankie can easily claim that they are indeed investigating why this didn't work and that they were not convinced they had optimized the conditions yet—indeed a plausible defense. Unfortunately, they know they were misrepresenting the data. The only ethical way to have proceeded here would have been to present the data that they had obtained, and at the end, make a statement that sends the message: "An exploration into why the allylic and benzylic substrates failed to give superior results despite decades of research demonstrating otherwise is under way." Of course, such an investigation really must be under way!

In Scenario 2, we will assume that the investigation into the benzylic and allylic really is under way. Without this assumption, Vlad and his mentor are behaving unethically. With this assumption, it should be equally clear that it is simply bad science. (Perhaps the case can be made that it is poor peer review in **both** cases, but that's a different discussion altogether covered later.) In Scenario 2, Vladomir made a safe assumption and generalization. There is nothing unethical about it even if in the long run, it turns out to be a bad, even stupid assumption. Bluntly put, it is not unethical to do something stupid! Based on his data, and his previous experiences, Vlad had every reason to draw the line that he did. It in fact is hard to even label this bad science. Maybe, they chose to publish too soon. Unfortunately, as Frankie's data shows (data that Vlad doesn't know about), Vlad was *wrong*.

Let's be clear on the distinction between these two cases. What Frankie **did** was wrong, but what Vladomir **thought** was wrong. If in every case of new science, we are not allowed to take for granted that certain previous trends will hold, then, really, what good is there in keeping track of anything? Making certain assumptions can be dangerous, but it's not unethical. In some cases, it comes back to burn you, but at the end of the day, assumptions are not unethical and being wrong is certainly not either. Science is largely dependent on standing on each other's shoulders. If we're not allowed to do that, no work can ever get done again. And finally, if Vlad being wrong *really* makes him unethical, at least he'll have company... I mean, really, who among us has never been wrong?

Something should also be said about the selective interpretation of data as well as it is as much bad science as it is bad ethics. An example of this is a relatively recent report[14] that over a lifetime, healthy individuals had a higher lifetime health expenditure than obese people or smokers. The study was done via a simulation taking various factors into account. However, the study appears to fail to consider issues like the quality of life for the individuals and appears to overlook the fact that healthier individuals inevitably contribute more positive things to society by way of missing less time at work and working for a longer number of years. Unfortunately, such selective interpretations can result when issues such as health care become politicized and corrupted to meet the needs of the organization behind the study. The reason why such selective interpretation can be considered both bad science and bad ethics is that science must be above such taint. Science is supposed to be completely objective, and when preconceived notions soil the interpretation, we are

---

[14] Van Baal, P. H. M. et al., *PLoS Medicine*, **2008**, 5, "Lifetime Medical Costs of Obesity: Prevention No Cure for Increasing Health Expenditure."

not following good science. Also, when we allow others to forcibly change our mind to suit their agenda, or when we deliberately only do such experiments that prove us right and not evaluate the validity of the claim, we're employing bad ethics, in addition to bad science.

NEW RESULTS THAT PROVE OLD RESULTS WRONG

The abovementioned hypothetical situations serve as an effective lead-in to a question that some readers may have at this point. This question is: "Has scientific misconduct occurred if someone is proven wrong?" Provided we assume the research was genuinely done and honestly reported, the answer to this question is, without doubt, always **no**. Science proves itself wrong on a routine basis. It is for this reason the answer to the related question of whether it is always bad science that is proven wrong is always NO. Instruments become more powerful and provide better resolution, allowing us to see three where we had previously seen only one. Genuine mistakes can be made in interpretation of data as well. Also, things are redefined, such as Pluto. Some of us grew up calling Pluto a planet, and now, it is not. Twenty years ago, science said Pluto was a planet. Today science says it is not. No bad science, no bad ethics, just scientific progress. Alternatively, there may be an unidentified decomposition or other influencing factors that cause a researcher to not really be observing what he or she thinks he or she is observing. An example where yields have been improved is in the development of what we now call Grignard reagents from the Barbier coupling reaction. In short, initial results (Barbier coupling reaction) involved mixing all the reagents together at once. The yields of these reactions were later shown by Grignard to dramatically improve if the organometallic reagent (specifically Mg) is prepared independently first. This, for certain, is not an example of bad ethics nor an example of bad science on Barbier's part. This is simply the natural progression of science. One researcher takes the preliminary work done by another and adds his or her own new insights, and this leads to the advancement of science.

In other cases, financial obstacles are overcome, and this leads to the production of more sensitive optics that allows astronomers to peer further into the universe than ever before. Does this mean that Galileo was a bad or unethical scientist for not creating the optics and using them on his first telescope? Nobody in their right mind would say yes. Similar cases to this last one can be mentioned *ad nausea*—anything from medical technologies to the light bulb. All are examples of the best of science.

One very common area that results are proven "wrong" is in the structure of natural products. Natural products often have enormous and complex chemical structures. Elucidating their full chemical structure is a monumental task that will certainly contain errors from time to time, even by the best researchers. Ultimately, these specific errors are often discovered when someone tries to synthesize the molecule and finds that the synthetic sample is not identical in spectroscopic or physical properties to the authentic sample. If he or she is then able to alter the synthesis to furnish a different chemical entity (often, only a small change such as the direction in which an atom or group of atoms is pointing is necessary) and this new chemical entity has identical spectroscopic and physical properties to the authentic sample, he or she has now corrected the initial structure assignment. This is something that happens often, and it is not an indication of bad ethics nor an indication of bad science.[15] In fact, I am tempted to argue that this is actually an example of good science. It demonstrates how science checks and corrects itself without the "help" of legislation or worse yet, congressional interference. It leaves everything in the hands of the people who know best—the experts.

THE WHISTLEBLOWER'S DILEMMA

Unfortunately, it is all too common that the whistleblower, that is, the individual (often a coworker of the violator) who called attention to the violation, suffers more serious repercussions than the violator. Of course, in cases of unfounded accusations, this is the appropriate response. In other cases, however, the whistleblower is fired and/or blackballed by the field. It should be made clear that peer reviewers do not usually see these repercussions. Occasionally, a researcher trying to reproduce the results may see some repercussions, although this is rare. These harsh repercussions are usually felt by those whistleblowers with whom the violator worked. This is a most unfortunate artifact of human nature, and it contributes to unchecked scientific misconduct and all aspects of society.

Three cases (and certainly not the only three, historically) where the whistleblower had negative repercussions inflicted upon them are Salvador Castro, a medical electronic engineer at Air-Shields Inc. in Pennsylvania; Margot O'Toole, a then postdoctoral research associate in the Imanishi-Kari lab at MIT; and Suzanne Stratton of the Carle Foundation Hospital in Illinois. The O'Toole case is discussed in greater detail from the context of fraud in Part B, with the focus here on the effects on O'Toole.

We will consider Castro's case first.[16] While working at Air-Shields Inc. (at the time based in Hatboro, Pennsylvania) in 1995, Castro identified a serious design flaw in one of the company's infant incubators. After reporting this flaw to his supervisor produced no changes in the design, Castro threatened to file a report with the U.S. Food and Drug Administration. He was then fired. Castro sued Air-Shields for wrongful termination, but the case still has not been resolved, in part because the company has changed hands more than once since firing him and since Pennsylvania employment laws at the time permitted employers to fire an employee without a reason.

Regarding the O'Toole situation,[17] many portions of the story are necessarily left out here, discussed in Part B. In short, work published by Imanishi-Kari and Baltimore was disputed by O'Toole. It was alleged by O'Toole that Imanish-Kari did not have data the paper claimed that she did and what's more, that data that was in hand was misrepresented. At one point, after reporting her discovery, O'Toole was allegedly told by Gene Brown, then Dean of Science at MIT, that she must either make a formal charge of fraud or drop the matter entirely, an allegation Brown later denied. Only after Imanish-Kari produced a compilation notebook during a congressional investigation did O'Toole finally allege fraud. A report from the office

---

[15] When searching Scifinder Scholar® for articles containing "structural revision" and the search was limited to articles published in 2009, 38 references were found! "Structural revision" is just one of the ways in which this is reported in the literature.

[16] http://spectrum.ieee.org/at-work/tech-careers/the-whistleblowers-dilemma last accessed 8/24/11.

[17] Judson, H. F. *The Great Betrayal: Fraud in Science*, Harcourt, Inc., **2004**, 191–243.

of Research Integrity ultimately found Imanish-Kari guilty, carrying a penalty of a 10-year ban from receiving federal grant money. She was also suspended from the faculty at Tufts, where she was employed at the time of the investigation. Imanish-Kari appealed and won her appeal, though the verdict was viewed by some as a "not proven" rather than a "not guilty." The decision went on to criticize O'Toole, claiming that she may have become too vested in the outcome of the investigation and that the extent to which she assisted the investigation may not have been appropriate. They even questioned the accuracy of her statements, despite her version of the events being the most consistent and unchanging of all the persons involved in this sordid affair. And, she was once referred to as incoherent by one of the researchers she brought her concerns to.

Suzanne Stratton, who holds a PhD in molecular biology, was at the time of this story the Vice President of Research at the Carle Foundation Hospital, a leader in cancer research. Stratton and the PI on a series of projects, Dr. Kendrith M. Rowland, Jr., himself, an oncologist, had a significant history of conflict. After 2 years of working at Carle, Stratton cited an outside audit that uncovered major deficiencies in 12 of 29 experiments investigated that were overseen by Rowland.[18] These deficiencies had the potential to harm patients or skew results (the details are beyond the scope of this book.) When Stratton voiced her concern to hospital administrators, they responded by firing her. Though they claimed her complaints were not related to Stratton's firing, one is free to interpret this otherwise "amazing coincidence" however they choose.

Wrapping Up

By now, you should be noticing a theme in why scientific misconduct occurs. In just about every case, it is used by the perpetrator to gain an edge or get ahead. This, in fact, is why all forms of cheating happen, whether it is a researcher claiming that a yield on a reaction is 89% when it is really 13% or an athlete using performance enhancing drugs or anything in between. Unfortunately, this is probably just human nature. We are a competitive breed, and, in all likelihood, we always will be. With that assumption in mind, the best that we can probably do is improving our ability to catch instances of scientific misconduct and our understanding of what scientific misconduct is so that we can avoid the truly accidental lapses in scientific integrity. Science, in some form, has been practiced for millennia. Over that time, we have become quite proficient at cleaning up after ourselves, even without legislative help. For sure, in cases where government money was used to fund the research stricken by an ethical violation, the government can and in fact should get involved somehow. However, there will always exist the international nature of science that makes it very difficult for any one government to "rule" on ethical violations. In fact, no government should be "ruling" on ethical violations. This should always be left to experts. What the governments can and should do however is to withhold federal funding to those who have been convicted by their peers. Scientific misconduct is also significantly different from bad science. The difference must be boiled down to acting wrong and being wrong, respectively.

---

[18] http://www.nytimes.com/2009/10/23/business/23carle.html?pagewanted=all, last accessed 6/24/18.

2 What Are the Penalties for Committing Research Misconduct?

The penalties for scientific misconduct vary from instance to instance. In some cases, the extent of the penalty is that the researcher's reputation is forever tainted, such as with Watson and Crick whom some people will never forgive for the controversy that enveloped them and Rosalind Franklin after they solved the structure of DNA.[1] Note, however, no formal charges were ever brought against the two. In more official penalties, sanctions would possibly be levied upon the researcher or the researcher may even be terminated.

At this point, it would probably be most instructive to discuss examples of real-life cases, starting with a case of plagiarism. In 2006, George Wagner wrote a letter to the editor of *Chemical and Engineering News* describing an experience he recently had in reading an article from a group of Chinese researchers. He wrote that he found some parts of the introduction to the paper in question to be "hauntingly familiar" and, upon a brief investigation, discovered that parts of it were reprinted virtually word for word from a paper that he had previously published. Wagner promptly contacted the editor of the journal, and the editor requested of the authors that an erratum and apology be written at once. An editor-in-chief of the journal that published the manuscript, *Inorganic Chemistry*, commented that "when it comes to copying boilerplate-type text for an article, 'for many Chinese authors, it's non-offensive'." Of course, in Western culture, this is a completely inexcusable offense, and furthermore, it is not tolerated by the science community at large. This does, however, bring us to an imbalanced set of consequences. Without doubt, there would have been at least *some* disciplinary action taken against a Western scientist by their institution had they committed this same offense, in addition to the mandate by the editor of the journal. No such penalties were reported here. As more than a decade has passed since this case, it is hard to say if there would still be virtually no penalties for Chinese researchers at their China-based institution.

When *fabrication* of data is discovered, the penalty is very severe. The offending researcher may be fired, be denied a degree (if they are a student), or even have funding revoked. Furthermore, the paper(s) containing the fabricated data would certainly be retracted so that nobody else falls victim to this crime on science, and significant embarrassment befalls the journal, the researcher(s), and the associated university or company.

---

[1] White, M. Rivals: *Conflict as the Fuel of Science*. Vintage Books **2002**, 231–273. Watson, J. D. *The Double Helix: A Personal Account of the Discovery of the Structure of DNA*. Simon & Schuster New York Touchstone Books, **1996**.

The case of Woo Suk Hwang demonstrates the fact that not only the researcher but also the field can potentially suffer severe repercussions when fraud occurs. Hwang was found to have fabricated data that contributed to two separate papers in *Science*. His research involved cloning and human embryonic stem cells. This is noteworthy because this field of research already had at the time shaky (at best) public support, at least in the United States. Any negative attention (and THIS certainly constitutes negative attention) could potentially bring funding and public support to a halt. To date, this (fortunately) has not happened. For his actions, Hwang nearly did time in prison. Not only was the data faked (Hwang admitted as much, but claimed he was duped by a colleague), but Hwang was found guilty of fraud as well. He was found to have accepted approximately $2 billion in private funds under false pretenses, and he was accused of embezzling approximately $800K and purchasing human eggs for research, a violation of contemporary South Korean bioethics laws. He was spared a jail sentence if he "stayed out of trouble" for 3 years, but then was fired from Seoul National University, and the South Korean government stripped him of his rights to conduct stem cell research, and *Science* naturally retracted both of his disputed papers.

The story, however, indirectly continues. Protocols established in South Korea in response to this debacle were employed in 2008 to help root out another case of fraud by Tae Kook Kim and colleagues who fabricated data during their investigation of a screening technology that will allow them to identify drug targets.[2] The investigators found that the data had been both fabricated and misrepresented by the authors of the two *Science* papers. This demonstrates how far-reaching, even indirectly, the repercussions of scientific misconduct can be. If the new protocols had not been in place, Kim and his coworkers may have been able to get away with their fraud for a longer time. This, naturally, would have caused greater harm to science.

A short aside is important here. That both incidents involved the journal *Science* is not a condemnation of the quality of this journal. *Science* remains one of the top two or three journals in the realm of scholarly scientific publications. As a result, publication in this outstanding journal remains one of the highest, if not **the** highest prize in scientific publication. One of the outcomes of this is that work in *Science* is some of the most heavily scrutinized work meaning that genuine errors are more likely to be uncovered than if the work is published in some obscure journals. This means that work is retracted not only for scientific misconduct but also for reasons of good science. Recently, the editors in chief for the journals *Infection and Immunity* and *mBio* published a report[3] described in *Biotechniques.com*[4] generating a "retraction index." They found that journals with a higher impact factor (such as *Science*) had a higher retraction index as well. Their study found that the top four retraction indices went to the *New England Journal of Medicine, Science, Cell,* and *Nature*. These four journals are four of the top five journals, with respect to impact factor, with *Lancet* being the fifth journal. The authors, while lauding the corrective nature

[2] www.sciencemag.org/news/2008/03/south-korean-researcher-suspended-over-charges-scientific-misconduct, last checked 6/24/18.

[3] Feng, F. C. and Casadevall, A. *Infect. Immun.*, **2011**, *79*, 3855–3859.

[4] Pun, S. *Biotechniques.com*, **9/13/11**, Higher Impact Factor, Higher Retraction Frequency.

of science of the study, go on to point out several of the consequences of retracted articles, especially for scientific misconduct reasons that are also discussed in this book:

1. The diversion of scientists down unproductive lines of research
2. The unfair distribution of scientific resources
3. Inappropriate medical treatment for patients
4. The erosion of public confidence in science
5. The erosion of public financial support for science

They also point out that the damage that even the relatively small number of retractions can do is disproportionate to relatively low number of retractions that occur.

The editors of *Chemistry of Materials* wrote a letter to *Chemical and Engineering News* in 2005 to describe an incident they had.[5] This case dealt with duplicate submissions and results that were too similar. Specifically, a paper in *Chemistry of Materials* was found to be essentially a duplicate of a paper in another journal. No specific details were provided by the editors in their letter, but they described their decision to take fairly (and deservedly) severe action and not only withdrew the paper in question from their web edition but also posed a 1–3-year ban on publishing in the journal on the authors. The editors go on in their letter to mention other penalties they would consider levying on perpetrators of scientific misconduct to include notifying their reviewers and the editors of previous journals a manuscript was found to be submitted to.

Recently, a former Duke Associate Professor, Anil Potti's **2006** paper in *Blood*[6] was retracted amid irreproducible results.[7] This article, though only published in 2006, had already been cited 24 times by the time of the retraction, according to the ISI Web of Knowledge. During an Institute of Medicine hearing, the Duke Vice Chancellor for Clinical Research, Rob Califf, testified that the University is nearly done with its investigation. Califf anticipates that of the 40 papers coauthored by Potti during his tenure at Duke, more than half will be either fully or partially retracted, a staggering number of retractions to say the least. Prior to the *Blood* **2006** retraction, four other Potti papers[8] had already been retracted. The investigation by Duke also discovered that Potti had lied on a grant application, claiming he was a Rhodes Scholarship recipient while he was not. Potti resigned from Duke amid the investigation while on paid administrative leave and currently holds a position at the Coastal Cancer Center in South Carolina. In 2015, Potti was found guilty of research misconduct by federal officials.[9]

---

[5] https://cen.acs.org/articles/83/i26/Ethics-scientific-publication.html, last checked 6/24/18.

[6] Potti, A. A.; Bild, H. K.; Dressman, D. A.; Lewis, J. R.; Ortel, T. L. *Blood*, **2006**, *107*, 1391–1396.

[7] Bialeck, M. *Biotechniques.com*, **9/2/11**, Five retractions and counting, last accessed 9/20/11.

[8] Bonnefoi, H. et al. *Lancet Oncol.*, **2007**, *12*, 1071–1078; Hsu, D. S. et al. *J. Clin. Oncol.*, **2007**, *25*, 4350–4357; Potti, A.; Dressman, H. K.; Bild, A.; Riedel, R. F.; Chang, G.; Sayer, R.; Cragun, J.; Cottrill, H.; Kelley, M. J.; Petersen, R.; Harpole, D. Marks, J.; Berchuck A.; Ginsburg, G. S.; Febbo, P.; Lancaster, J.; Nevins JR. *Nat. Med.*, **2006**, *4*, 889; Potti A.; Mukherjee S.; Petersen R.; Dressman H. K.; Bild A.; Koontz J.; Kratzke R.; Watson MA.; Kelley M.; Ginsburg G. S.; West M.; Harpole D. H. Jr.; Nevins J. R.; *N. Engl. J. Med.*, **2006**, *355* 570–580.

[9] www.sciencemag.org/news/2015/11/potti-found-guilty-research-misconduct, last accessed 6/24/18.

It is not only academic institutions but also governmental research institutions that suffer from scientific misconduct. Starting in late 2003, the NIH was involved in a congressional investigation into conflicts of interest committed by institute scientists. By the end of the cooperative investigation by the NIH and a House committee, 44 scientists were identified as having violated NIH policies or rules governing conflicts of interest. Of whom, 36 were referred for administrative action, 8 were no longer NIH employees by the end of the investigation, and 9 were referred to the Department of Health and Human Services Office of Inspector General for further investigation. Additionally, the NIH identified 22 further cases that the House Committee did not identify and initiated an investigation of these scientists. In this case, the NIH and House Committee worked together productively to address serious concerns that challenged the integrity of the NIH. All the individuals involved deserve high praise for working together harmoniously and efficiently.

The main violation that occurred was that scientists were failing to disclose potential conflicts of interest. The transgression of these misbehaving relative few (the NIH employs thousands of scientists) brought severe repercussions down on everyone. The new rules that were borne out of the investigation can be summarized as follows:

1. NIH employees are barred from participating in paid or unpaid activities with pharmaceutical or biotech companies, health-care providers, health insurers, trade and professional organizations, and higher education or research institutions *that hold or are applying for agency grants*.
2. Employees of the NIH are restricted from owning more than $15K in pharmaceutical, biotech, or related company stocks.
3. Senior level employees of the NIH were prohibited from holding any stock in the abovementioned sectors.
4. No employee can accept an award more than $200 unless it is a major scientific award (i.e., Nobel Prize).

Repercussions (probably unintended ones) were felt immediately. For example, James F. Battel, at the time director of the National Institute on Deafness and Other Communication Disorders, understandably felt compelled to retire since significant family investment would have to be divested, bringing about a significant tax burden to maintain compliance with rule 3. Also, a survey revealed that ~40% of tenure or tenure-track scientists were or had considered searching for a job outside the NIH because of the new regulations. In late 2006, Pearson Sunderland III was sentenced to 2 years' probation under the new rules for activities he committed from 1998 to 2004. The maximum penalty for Sunderland's actions was up to 1 year in jail and a fine of $100K. This would have been a significant penalty, considering that most people consider a crime had certainly not taken place. This, however, is the importance of trust in science. If trust is violated, especially when federal dollars are influenced, the penalty must be severe to protect the trust.

Another case involves a very modern form of data manipulation. The *Journal of Biological Chemistry* has detected fraud in digitally altered photos.[10] For example,

---

[10] www.jbc.org/site/misc/photoshop.xhtml, last accessed 4/19/11.

they found cases of reuse of control images within a single publication without noting the repetition, figures from one manuscript being used in another for *new* purposes, and digital removal of contaminating bands from gel patterns. The consequences in each of these cases were the rejection or withdrawal of the papers in question and the notification of the authors' institution (a common and standard penalty). Most people would immediately agree that the third of these offenses is indeed an instance of scientific misconduct. The first two, however, may not be so obvious to a more inexperienced researcher. Consequently, a discussion of why these two are examples of scientific misconduct is appropriate here. First, neglecting to specifically note that a control image is being reused in a series of studies, even within the same paper, leaves out an important piece of data that puts the results into context. While it can certainly be argued that it may have been an honest mistake, or perhaps even bad science rather than bad ethics, these incidents must be avoided whether classified as bad science or bad ethics, especially since it is usually trivial to add a single sentence that makes known a control image's reuse. For certain, if the omission of such a statement influences the interpretation of the data, it is a violation of scientific ethics and falsification of data. It can also be argued that it is just logical to reuse the same figure throughout a single publication. For sure, this is a good argument, but since the matter can be resolved with the addition of a single sentence, it is better to just be clear and tell this to your readers. The second case is potentially even more confusing. For certain, a researcher can mention and/or reuse prior results, such as an image, in a subsequent paper, provided that (1) enough new research is also reported that either builds upon the research previously reported or shows it in a new light and (2) it is properly referenced.[11] We must be careful here to notice what, specifically, the offense was. The offense was said to be that the figures from one paper were being used in another for *new* purposes. This is a very important distinction, and it is this that makes this action scientific misconduct.

In other cases, such as the Baltimore case discussed in detail later in this book, a researcher may be banned from applying for federally funded grants or cut out of funded grants if found guilty of scientific misconduct. As will be seen when this case is discussed later, the researcher in question was then relieved of this restriction, not because of what was considered by some to be full absolution of wrongdoing but because of what were declared to be inappropriate actions on the part of the investigators. The fact remains, however, that fabrication of data is dealt with severely. To ban a researcher from a family of grants is to effectively blackball him from the scientific community, since without grant funding, there is quite literally no way for a researcher to perform science. It is certainly comparable to a (professional) death sentence.

The court of public opinion and the "penalties" it imposes are also one that is tremendously powerful. For example, in early 2009,[12] Carl Djerassi wrote a letter to *Chemical and Engineering News* objecting to a lack of thorough crediting by Trost and Dong of early Bryostatin work. The action that Djerassi took great exception to

---

[11] This is not the same as using a stock file of chemical structures you prepared to increase efficiently later. Here, we are considering the reuse of a *result* in image form.

[12] Djerassi, C. *Chem. Eng. News*, **26 January 2009**.

was Trost and Dong's decision to cite a general review of earlier work, written by authors other than the researcher who isolated and first investigated this important class of natural products. These issues can cause divisions in the science community. Although it can certainly be argued that a diligent reader can certainly find the original work using the references provided by Trost and Dong, a fair statement for sure, it is simply not "right." It is always most appropriate to reference the original work rather than a review when crediting discoveries.

The Royal Society of Chemistry reserves the right to expel violators of responsible conduct of research from their society. In instances where publication in a particular journal, application to a grant, attendance at meetings, or admission into professional networks is at least in part influenced by membership in a particular society, this can be a particularly stringent (but still appropriate) penalty.

Other consequences are perhaps better explained using hypothetical situations. For example, imagine that you are visiting graduate schools that you have been accepted to and are trying to decide which to attend. During one of your visits, you speak to a student who left the group that you are most interested in joining at this university in his or her third year, a very risky thing to do so late in one's graduate career. After asking them why, he or she informs you that he or she left because of a dispute in authorship of a paper—he or she felt that he or she did enough work to be a coauthor on the paper, but one of his or her junior lab mates argued to the advisor against it and the advisor decided **not** to include him or her as a coauthor on the paper based upon the argument. This should quite naturally give you cause for concern. It would not be unreasonable for you to rethink whether you would be best served joining this research group. This is yet another reason that one must be very careful when deciding whom to include or not include as a coauthor on a paper. Being very demanding on this important issue may discourage otherwise excellent people (in this case, you!) from working with them, either as collaborators or as students.

Breaches of confidentiality could come along with extremely severe penalties. In mid-2010, Ke-Xue Huang, formerly of Dow Agrosciences, was arrested under the federal Economic Espionage Act of 1996. He was accused of sending confidential information about insecticides to collaborators while authoring a (legitimate) review article in 2009. Huang pled guilty in late 2011 and was sentenced to 7 years in prison.[13] Liu Wen, also of Dow Chemical, was found guilty of conspiring to steal company secrets. Wen was found to have paid about $50K in bribes to a Dow employee to supply materials about how Dow produced certain polymers.[14]

The fallout from scientific misconduct does not have to only affect the perpetrator of the misconduct. The graduate students or other laboratory associates feel the aftermath of scientific misconduct when the principal investigator is found to have committed scientific misconduct. Take, for example, the case of Elizabeth Goodwin, former faculty member at the University of Wisconsin (UW).[15] Goodwin was found

---

[13] www.bbc.com/news/business-16297237, last accessed 6/24/18.
[14] www.nytimes.com/2011/2/08/business/global/08bribe.html, last accessed 8/2/11.
[15] www.uwalumni.com/home/alumniandfriends/onwisconsin/owspring2008/worms.aspx, last accessed 9/22/11.

to have fabricated data on a grant application. This finding was made by the students working in her lab. After Goodwin resigned amid the controversy, only two of the seven researchers (two students) working in her lab at the time were able to find other positions within the department at UW. The other four students and research specialist chose to depart UW, with one of these five completely changing careers to enter law school instead. The investigation continued even after Goodwin left because Bill Mellon, then associate dean of the graduate school and head of UW's research compliance, felt that the university had an obligation to investigate the charges. Part of his contention was that the misconduct involved at least two federal grant applications, a renewal, and an application for new funding. Mellon went on to point out that the university must show the federal government that they are serious about the honesty of the scientists employed at UW. Curiously, there has been no question of the integrity of the data Goodwin had published in various manuscripts. It appears that the foul committed was "only" fabricating data on a grant renewal application.

This particular case is noteworthy because it was primarily a group of graduate students in the lab who discovered the foul committed by their research mentor. The research specialist also played a role, but the role of the graduate students should not be scarred by this. These graduate students were taking an enormous risk—one that may have ended their careers before they even really started. These were *students* still on their way to their graduate degrees. They had the courage and sense of what was right to bring this story to light is a testament to the things that are right about science, and they should absolutely be lauded for their actions. It often takes considerable courage to even disagree with one's research mentor on a scientific topic. It is an entirely more complicated and awkward matter to take up claims of scientific misconduct against one's mentor.

With respect to repeated publications, this is one ethical violation that has a built-in penalty. For sure, as discussed earlier, more publications often mean a better career, but the rest of us in the field are not stupid. When we see multiple publications with many similar titles, we know what's happening, and yet people do it anyway. Furthermore, if one massive paper is instead divided into three or four smaller papers, for sure, the publication record is thickened, but it is of great prestige to have your article cited more times. The more spread thin the work, the less any one of the publications would get cited, in principle. Hence, it may be detrimental to the offending researcher in a more indirect way, it is certainly detrimental to the perception of the overall quality of the work in the long run, and it is also self-plagiarism as discussed before and a clear ethical violation.

In another case, a researcher has been incarcerated for his or her scientific misconduct—Eric Poehlman,[16] former professor at the University of Vermont (UVM). Poehlman was found to have fabricated data for at least ten manuscripts submitted to different journals.[17] He presented faked data during seminars. He was also accused of reporting false data in a funded grant application, a federal crime with a maximum penalty of 5 years in federal prison. During his defense, Poehlman lied

[16] www.nytimes.com/2006/10/22/magazine/22sciencefraud.html?pagewanted=1&_r=1, last accessed 9/22/11.
[17] Sox, H. C.; Rennie, D. *Ann. Intern. Med.*, **2006**, *144*, 609–613.

under oath but did not reportedly face charges of perjury. As the case built against him, Poehlman left UVM to take a position at the University of Montreal. With pressure mounting and the threat of prison hanging over his head, Poehlman finally offered his full cooperation, ultimately pleading guilty to the charge of making fraudulent claims in a grant. He received a sentence of 1 year and 1 day in federal prison plus 2 years' probation. He was also ordered to pay nearly $200K in restitution. He was further banned *forever* from receiving public money. Poehlman's research involved searching for a correlation between weight gain and menopause, among other things. At his sentencing, Poehlman had some choice words about the scientific establishment:

> I had placed myself, in all honesty, in a situation, in an academic position which the amount of grants that you held basically determined one's self-worth, everything flowed from that. With that grant, I could pay people's salaries, which I was very, very concerned about. I take full responsibility for the type of position that I had that was so grant-dependent. But it created a maladaptive behavior pattern. I was on a treadmill and I couldn't get off.

Such incentives for scientific misconduct are mentioned elsewhere in this book. This, of course, does not justify them, nor should we feel pity on those who succumb to the pressures Erick Poehlman waxes about. The fact remains, however, that such incentives are in fact real, and they drive the foolish and unethical choices that a few (and truly, just a few) scientists make.

HUMAN OR ANIMAL SUBJECTS

Regarding human and animal subjects, extreme care must be taken to ensure the proper care and oversight for the well-being of these animals. This care, with respect to human and animal subjects, is covered in more detail later. Part of why so much care must be taken is not only because *life* itself is at stake but also because there are severe, official/financial/federally enforced penalties and sociological penalties for the errors that can happen. Animal rights activists are among the most passionate and vocal for their cause groups in all of society. Therefore, drawing their ire is something worth avoiding. Although the U.S. Department of Agriculture (USDA) inspects the research facilities in this country that work with warm-blooded animals each year to ensure the humane treatment of the animals, it does not dictate research protocols. Instead, the agency forces the facilities to establish their own ethics oversight committees that guide their actions and decisions on the Animal Welfare Act.

A recent example is the citation of two Harvard animal research labs by the USDA.[18] This facility was cited after the second primate death in as many years. During an investigation, Paula S. Gladue, USDA Veterinary Medical Officer, found several violations of the Animal Welfare Act. These violations included unsanitary conditions in the operating room, unqualified staff, and dirty (animal) housing facilities at the Boston campus. Regarding the unqualified staff charge, an anesthetist

---

[18] Pun, S. *Biotechniques.com*, **9/9/11**, Harvard Animal Research Facilities Cited.

overdosed a nonhuman primate with an anesthetic agent during a surgical procedure and the animal subsequently died due to kidney failure. Harvard responded to this by retraining the relevant staff in the proper use of anesthetics to try to prevent such an incident in the future. Michael Burdkie, executive director of Stop Animal Exploitation Now! had particularly harsh comments, saying "When there's an incident like this, when a staff has overdosed an animal causing death, it's clear that the staff is unqualified and should not be allowed to use animals in the future." Such comments are almost for sure, overly harsh. It is indeed a tragic mistake, but, however unfortunate, mistakes happen. Unless the worker was deliberate in his or her actions, or grossly incompetent, such words are indeed too harsh. However, this is exactly the concern mentioned earlier—animal rights activists are among the most zealous of all activists.

The room cleanliness violations stemmed from the finding that the operating room had rust and peeling paint. As if that wasn't enough, primate chairs were found to be covered in residue and fecal matter. The facility was given a few weeks' time to correct these violations to avoid further action being taken by the USDA.

Harvard's New England Primate Research Center was also found to have committed violations of the Animal Welfare Act by conducting procedures that were not approved by the facilities' Institutional Animal Care and Use committee. This may mean that one procedure was approved by the oversight committee but another, different, and non-approved procedure was used instead. Alternatively, the researchers may have just begun work with no approval at all. Both of these scenarios are serious violations.

Previously, at the New England site, the USDA found other violations as well. Specifically, a cage was cleaned in a mechanical cage washer, while it still contained a dead nonhuman primate. It is unclear from the report why this deceased animal was left in the cage at the time of the cleaning. Following these events, Mimi Sanada, an apprentice caregiver at Jungle Friends Primate Sanctuary, offered a very level-headed but firm response, saying "These types of incidents should not happen, and a warning or a fee is not sufficient. The fact that another grave mistake was made by the same facility following this past incident is indicative of a need for a much more avid and frequent inspection to ensure all of the animals' safety." Ms. Sanada's point of this not being the first offense is particularly important. If these lapses had occurred at different research facilities across the nation, each incident would be no less tragic, but it would be less indicative of a culture problem at any one facility. That each of these incidents happened at facilities under one administrative body suggests that there *may* be a culture that endorses a shirking of the rules or at the very least one where the training is insufficient.

WRAPPING UP

As can be seen from the previous discussions, the penalties for scientific misconduct vary significantly. There are multiple factors influencing the punishments handed down including the severity and scope of the misconduct, the culture of the "jury" (i.e., Europe vs. U.S. vs. Asia) and whether federal monies were involved. This will likely always be the case and that is probably OK. By and large, although sociological penalties may vary, the scientific ones do not. In nearly all cases (likely all, in the absence of exoneration), the researchers' work will never be wholly trusted again, their fraudulent or plagiarized work will be retracted, and they may even find themselves "blacklisted" by select journals, funding agencies, or both. Many of these penalties, especially the journal blacklisting, is a punishment exacted by the *international* scientific community. Such punishments are not unfairly likened to excommunication.

3 Peer Review
What Is It and What Is Its Role in Scientific Misconduct?

Peer review is under most circumstances unpaid and volunteer. Editors and/or associate editors are generally free to choose anyone to serve as a reviewer. Some, in fact a majority of, journals permit the contributing author to recommend reviewers or even request particular individuals **not** be reviewers. The editor, in any case, is wise to choose a reviewer who is an expert in the same field as the paper or manuscript (for the sake of this book, the word manuscript is used in nearly all cases but here both refer to the same piece of scholarly work covers). Ordinarily, and especially in grant reviews, there are lists of colleagues or otherwise close associates (the National Science Foundation (NSF) calls them collaborations or other affiliations). These persons are disqualified from review to avoid a conflict of interest. The major flaw in this approach is that such professional or personal lists are self-reporting. Authors are left on the honor system to do so truthfully. Also, some fields, especially the newer ones, are very small. This virtually ensures that the entire expert pool is intertwined and holds some manner of conflict of interest, either in support of or in opposition to (competition with) one another.

Although it is not without its detractors and certainly not flawless, peer review remains an integral part to the current modus operandi of scientific publication and scientific grants. A deeper discussion of the ethics of peer review will come later, but first, it will be helpful to discuss what it is and how it is done.

While the very nature of science places almost all (interesting enough to be cared about) work under constant peer review, such is not what is typically meant when talking about peer review, although there are calls for it to become this, which we'll also cover later. Instead, peer review is the process of experts evaluating a manuscript's fitness for publication. After a manuscript is submitted to a journal, the editor will perform an initial screen of the paper to verify it falls within the range of the subject area of the paper. The editor can, at this point, reject the paper without review. This may be because he or she feels that it is not likely to be important and there are other higher-pressing submissions or that the manuscript falls outside the scope of the journal. As an extreme example of the latter, a mathematician is unlikely to submit a paper on some new calculus theorem to the *Journal of Organic Chemistry*. A historian is equally unlikely to find a captive audience for their paper on the U.S. Civil War in a computer science journal. Once the manuscript passes this initial screen, the peer reviewers then also will comment on the appropriateness of the manuscript for the journal. Although it is a small aspect of peer review, this acts to help ensure that articles of a field are all in the same space, at least to some great extent. This allows for much easier searching on the part of the greater community. High-profile journals such as *Nature* or *Science*, which tend to publish papers from a very broad range of topics, also exist, however.

Peer review also verifies the science in at least two ways: that the experiments reported were appropriate to solve the research problem at hand and that the data obtained is consistent with the experiments completed. Peer review also validates that the conclusions are logically drawn from the evidence/results collected from the experiments. That's not to say that peer review stamps the work as "right," but more on that point when we talk about what peer review isn't.

Peer review also serves at least some measure of editor. For example, if some portion of the manuscript is structured so poorly that the results or conclusion are being obfuscated, the reviewers can make suggestions about the manuscript needing to be edited. The reviewer may also suggest that additional experiments need to be run to further strengthen the argument of the authors. They also scrutinize the introduction portion where authors typically discuss the prior work done and where their work and/or hypotheses fit into or further develop the prior work.

Armed with this admittedly short list on what peer review is, a coverage of what it is not is appropriate. In no particular order, peer review is not

- An editorial service
- A fact-checker
- An unethical behavior detector

Considering these in turn, peer review is not an editorial service. It is not appropriate to submit a rough draft of a manuscript with the intention of utilizing the peer-review process to clean up the paper. This sort of rough draft is for friends or other trusted colleagues. That's not to say that the process doesn't provide some such feedback, but to have it be **the** method of cleaning up your manuscript is inappropriate. Peer review is done on a volunteer and unpaid basis. Submitting rough drafts wastes the already heavily burdened system's resources. Similarly, it is inappropriate to submit a manuscript to one journal with the hopes of getting quality feedback to improve the paper only to withdraw the submission, revise the paper based upon the feedback received, and eventually submit the (improved) manuscript to a higher profile journal. This, like the editorial feedback on a rough draft, wastes the time and resources of the volunteer peer reviewers. It also a violation of the good faith that manuscripts are submitted with. To do this is to knowingly submit something that you know you do not intend to follow through on. It is deceptive and wrong.

Also, let's be careful to not overstate the meaning of the evaluation of the science in the manuscript; it (the evaluation) is not a fact-checker. Although peer reviewers are certainly experts in the field, their function in the review process is merely to verify that there is congruence between the aim(s), the experiment(s), the result(s), and the conclusion(s). This in **no way** certifies that the experiments actually happened as written, nor does it insinuate that future experiments will not lead to different, even contrary conclusions. This last point is especially hard for the nonscience public to understand. The self-correcting nature of science is something scientists would be very wise to better articulate in the future and is why we should be very careful when we publicly communicate with popular, lay audiences. Circling back for a moment to how peer review does not verify that any experiments or set of experiments actually happened, peer review is not equipped to detect most forms of science misconduct.

With what peer review is and isn't now in our pocket, let's move on to how it is done, limiting our coverage to manuscripts in peer-reviewed journals and ignoring grants and books. In a peer-reviewed journal, the submission is generally sent to two to three reviewers by either the editor-in-chief or one of the associate editors. The reviewers then are to, unless given explicit permission, confidentially review the document, in particular assessing the scientific merit of the manuscript. It is the reviewer's job to evaluate if the conclusions reached are consistent with the data gathered and furthermore that the experiments were appropriate to yield such data and to solve the problem at hand. The peer reviewer ought to also comment on any alternative interpretations of the data and if necessary suggest additional experiments for the authors to do in order to arrive at a most (hopefully) correct conclusion. One must be careful, however, to not extend the work beyond the scope of the manuscript. Thus, it is a best practice to differentiate between comments to make the paper stronger and potential future work. Although reviewers will inevitably be distracted by very poor writing, they should take pains to avoid editorial comments.

There are even several guidelines regarding how peer review ought to be done, one notable one is from the Committee on Publication Ethics.[1] This document provides several dos and don'ts regarding peer review.

As a brief summary, the first part of peer review should be obvious: thoroughly read the manuscript and all the associated materials. If a reviewer feels that some material is missing, he or she should contact the editor, *not the authors*. Under no circumstances is it OK to contact any authors without the explicit permission of the editor. Once the materials are read, the review can proceed. On the way, however, confidentially must be maintained as follows:

- Do not use information for your and any or any of your associates' advantage or disadvantage (of a competitor).
- Do not involve anyone else in the review without the permission of the editor.
- Include the name of anyone who is given permission to assist.

Additionally, a reviewer must remain nonbiased and be free of competing (conflicts of) interest. A bias (however, unfortunate) may come from issues related to gender, nationality, religion, political beliefs, or financial interests, just to name a few. If any such situation were to present itself, the reviewer would be obligated to contact the editor at once. If, on the other hand, any one of them exists at the time of invitation to review, the reviewer is to make this known and decline the invitation. Also, a peer reviewer may serve as an unofficial first line of defense against misconduct, as stated earlier. As experts in the field, a peer reviewer is likely to recognize uncited text as plagiarized. Also, such reviewers may be more adept at identifying when a result is

---

[1] COPE council. Ethical guidelines for peer reviewers. September 2017. www.publiationethics.org.

too good to be true. In any such case, the reviewer should report his or her suspicions to the journal editor and not launch an investigation of their own. If called upon to provide further information or evidence by the journal, they should do so. A reviewer may also coincidentally be asked to review the same manuscript after rejection and subsequent submission to different journals. In such a case, there is no way to know if any changes have been made. The entire manuscript must be thoroughly reread. It may be appropriate, but only with the first journal's permission, to share this initial review with the second journal.

In preparing the actual report, some journals have specific rubrics/scorecards or other formats to use. If a journal you are reviewing for does, use them. The feedback provided should be fair and professional with the aim of strengthening the manuscript. This does **not** mean "do not be critical." Criticism is important, even essential to the process, but it focuses on constructive criticism. Most journals will ask reviewers to recommend action (e.g., accept, revise, reject). Whatever the recommendation is, it must be consistent with the feedback. That is, it would be inappropriate to laud the paper and then recommend rejection. The two are incongruent with each other. Journals also often have separate fields for comments—comments to the author (which the editor sees) and comments to the editor only. This should not be abused and taken as an opportunity to unfairly criticize the author. It may be best to adopt a policy of not saying anything to the editor that you would object to the author seeing. Suspicions of research misconduct should not be laid out here. If a reviewer suspected misconduct, it should have been addressed earlier in the process. Finally, confidentiality obligations do not end with the completion of the review process. Even after a paper is published, a reviewer ought not to speak of his or her role without permission from the editor.

Another point of collegiality worth noting is that a reviewer must keep in mind that research is a global enterprise now. This means that the author may not be writing in a language that is his or her primary language. Excessively harsh criticism of the writing is therefore not appropriate. This does not mean that poor writing should be overlooked, especially writing that is so poor and the meaning of the results is obfuscated or changed. It just means that collegiality should trump being a "grammar snob."

Although peer review is perfectly capable of detecting some blatant instances of scientific misconduct, it is certainly not trivial. It is also not really the primary goal of peer review. For it to be, deceit would be assumed, rather than truth and with that change, the whole scientific enterprise would collapse. Why the hyperbole? Because with literally thousands of scientific journals being published every year and each journal publishing hundreds of if not over a thousand articles, the sheer volume of work that currently exists (especially when it is considered that reviewers are almost always unpaid volunteers) makes it impossible for reviewers to comment on not only the quality of the science and its presentation but also its validity. Even determining that the submission contains novel work is not always something that can be done easily. When it is considered that nearly all research is carried out by experts with years of very specific research experience, it is not practical for a reviewer to take up the task of also verifying the research especially if the work being reviewed is a study with years of data or results from a custom-built instrument. Verification of the research is usually left to the greater scientific community after the research is published. Typically, work that is important enough to be used or further developed has a large number of people who to try to use it.

Although peer review has at least an obligation to report ethical misconduct, it can certainly be argued that in some cases, the peer-review process in its current incarnation may encourage some of its occurrence or even be unethical itself. For example, it is not unheard of for a reviewer to reject a paper based upon one negative or aberrant data point. The impact that this has on science is that authors become hesitant to report all their data, or to, worse yet, alter their data in order to avoid having their paper rejected because of it. In cases such as this, when a reviewer is being unreasonable in his or her justification for rejecting a paper, the editor must take action and either opt to publish the paper over the reviewer's objections or send the paper to an additional reviewer for what would hopefully offer a more reasonable review. Do not misunderstand; I'm not blaming peer review for these forms of scientific misconduct. I'm merely arguing its typical modus operandi gives the author reasons to be concerned. But it is always the cheater's fault.

These review tactics do not go without their consequences for the reviewer, of course. A conscientious editor, for example, would take note of which reviewers behave in this way and avoid giving such reviewers as many papers to review in the future, or at least until they prove that they have changed their review tactics or that this was a one-time event.

Part of why reviewers can get away with such behavior is that reviews for most journals are anonymous (at least from the author's point of view). This means that a reviewer need not fear for retribution from the author for an unreasonably negative review. For sure, the editor knows who reviewed each paper, but the editor may not know that there are personal or professional conflicts that may have influenced the review. Remember that science is based on trust. Thus, an exceedingly negative review may not *appear* unreasonable to the editor; it may simply appear *diligent*. I personally know of someone who has written extremely long and even picky reviews from every angle: science and the writing. This person is one of the most stand-up people that I know, and I can honestly say that he or she holds himself or herself to the same standards when he or she writes his or her own papers.

Despite the difficulties with anonymous peer review, it is likely the best option, though the tides may be changing. This is because it allows for exactly what was just discussed—it allows for a *diligent* reviewer to harshly criticize a paper deserving of such a critique without fear of reprisal from the authors. This check on science is essential to the progress of science. If reviewers were to be compelled to critique the manuscripts less harshly, for fear of reprisals, the science will inevitably become sloppy.

The argument can also be made that anonymous reviews facilitate the theft of data by the reviewer. Coming back to the importance of trust in science, this argument holds no water. Even with the anonymous review system currently employed, theft like this is exceedingly rare. And, if it does happen, the scientific community will ensure that this individual is never in position to do so again.

But can peer review prevent scientific misconduct? In some cases, the answer to this query is *absolutely,* but in others, it is *absolutely* **not**. Either way, trying to do so is very risky. Communication in science is very much based on trust. Employing this sort of tactic assumes misconduct. The peer-review process, although it usually does not do so presently, can, for example, be modified to include an additional reviewer whose sole responsibility is to check all the cited references to at least make sure all the references are cited properly. This may be smart anyway since it would also detect honest typographical errors in citing as well. The peer-review process could also be modified to where **all** journals would require statements from all the authors to give consent to publish; agree with the author order; and that no other deserving authors were omitted from the manuscript. Stealing someone else's work can also easily be detected by the peer-review process in cases where the previous work done is already published. This can easily be done by once again employing an additional reviewer (perhaps even the same reviewer checking citations) to perform a "novelty evaluation" where he or she searches for the research topic using a combination of scholarly search tools or some manner of plagiarism detector software, which catch all forms of written plagiarism. This also may mitigate the number of instances of intentionally neglecting previous work done, but the process is already quite good at detecting this. However, if the manuscripts are submitted simultaneously, nothing can uncover them other than the same person reviewing both articles.

Misrepresentation of previous work done *can* be caught, but it also can easily pass unnoticed even though experts in the field serve as the reviewers; they may not always be intimately familiar enough with the cited work to know if it is being represented fairly. To expect any reviewer to read every reference (even an extra reviewer) is completely unreasonable, the reviewers are almost always unpaid volunteers, after all.

The infractions the peer review is unable to identify and perhaps should not be expected to identify are Conflicts of Interest and Breaches of Confidentiality. In fact, it can easily and logically be reasoned that the responsibility of preventing these forms of scientific misconduct lies with the institutions that the researcher in question works for. Alternatively, a reviewer can certainly commit a conflict of interest violation or a breach of confidentiality violation. In a case like this, the responsibility still does not fall on the peer-review process to detect this infraction; instead, it is the editor's job.

The infractions that peer review is categorically unequipped to detect (and arguably to worst of the infractions) are falsification/fabrication of data and the closely related deliberate omission of conflicting data. These infractions are impossible to identify via peer review.

First, research is often very expensive. To fund the peer-review process checking, every result would increase the cost of research dramatically. It will likely not quite double it since the exploratory work needn't be repeated, just the final experiment but it would not be cheap. Also, science works largely on trust, and the peer reviewer has no choice but to believe the authors that the experiment as described was truly carried out and that the results reported are really the outcome of said experiments. That being said, the test of time will often ferret out falsified research. For sure, since this would typically be done by other researchers in the field, this can be argued as *extended* peer review doing so. Remember though that such is not traditionally what is commonly meant by peer review. Yet some journals like *Organic Synthesis* do check procedures, but there is no earthly way *all* journals could ever be set up like this. This level of fact-checking is the only way these violations can ever be found via peer review. The ironic part is that at least in the case of *Organic Synthesis*, the fact-checking is not done to detect scientific misconduct. Instead, it is done to independently verify a synthetic protocol viewed as being highly important for widespread use in the synthetic community. The reality (sometimes unfortunate reality) is that these forms of scientific misconduct are discovered by the general science community eventually. Sometimes, this happens at great expense to the researcher(s) who try to use or replicate the fabricated results, and this is, of course, unfortunate. But science is amazingly good at self-correction, and such violations are almost always eventually caught, and in cases of influential work, such violations **are** always caught. Data being omitted, on the other hand, cannot be caught by repeating the work that was done, since the reviewer is under the impression the work was not done, rather than omitted.

WEB 2.0 AND PEER REVIEW

Tides regarding peer review may be shifting, however. What they are shifting to is commonly referred to as Web 2.0. It effectively recognizes social media platforms such as blogs, Facebook, and webpages as viable publication venues. Those opposed to such a model point out that this "publishes" results before they can be peer reviewed. This also has the potential for publishing research with critical flaws or fabrications and/or falsifications. These arguments are countered by pointing out that such a publication model makes the peer-review process perpetual, and far more open and wider in breadth. Under such a system, every reader, particularly those who leave some manner of comment, becomes a peer reviewer. This very point is why research with critical flaws being published is a concern that doesn't hold water. Since most social media venues have room for adding comments and/or discussion, such critical flaws would be easily pointed out and maybe even resolved by a crowdsourced readership. Perhaps an intermediary step where publishers add a discussion or comment field to their articles (web versions, anyway) may be worth exploring. The *British Medical Journal* appears to already allow comments to be made on articles on its website. As of this writing, the *Journal of the American Chemical Society*, the flagship journal for the American Chemical Society does not allow comments on its web-based articles. The organic synthesis journal *SynLett* recently reported on its experience with what it called intelligent crowd reviewing.[2] Benjamin List, editor of *SynLett*, describes the journal's experiment as one where a large pool of reviewers was given 72 h to post or respond to anonymous comments made on submitted manuscripts on a protected online platform. Parallel to this novel form of manuscript review, conventional peer review was done on the same papers. List claims that taken together, the reviews were at least equal in attention to detail as reviews from conventional modes. In all ten of the papers used in this experiment, editorial decisions were made much faster (days rather than months) and that collectively the feedback provided was more comprehensive. So pleased is List and the rest of the editorial staff, that they are planning to make this the main process by which their manuscripts are reviewed.

---

[2] List, Benjamin, *Nature*, **2017**, 9.

REVISITING VLAD AND FRANK

Recall from Chapter 1 where we discussed the difference of bad *ethics* vs. bad *science*. Remember that Frank deliberately ignored data he had collected, and Vlad assumed that something he hadn't done (yet) would work based upon decades of data. It certainly would not have been inappropriate (although it would perhaps be draconian) for the peer reviewers to have rejected both manuscripts, insisting that the researchers perform the study on the allylic and benzylic systems and report the complete work in one paper. In this way, the peer-review process will be indirectly preventing misconduct by way of forcing the work (that is being omitted in Frank's case and under way in Vlad's case) to be performed and included in the paper. Essentially, by enforcing a high standard in the science, they inadvertently stopped Frank's misconduct.

CAN PEER REVIEWERS BE UNETHICAL?

Without doubt, the answer to this question is **yes**. First of all, we previously discussed in this chapter unfairly harsh reviews and stealing reviewed work. There is unfortunately no accountability with the current peer-review process since it is anonymous. Another instance where a peer reviewer can behave unethically is by somehow interrupting the publication process. If a reviewer is sent a publication of a competitor, and he or she allows the publication to sit, unreviewed (or without a review being submitted) for months while his or her own lab finishes competing work, he or she is abusing the system and committing scientific misconduct. Reviewers have an obligation to be timely and prompt, in addition to fair, thoughtful, and honest. I have been personally lamented to by a researcher who felt a reviewer or editor deliberately held up a manuscript so that it would not appear in the first annual issue of a high-quality journal. Why does being in the first issue matter? Major journals publicize the most cited paper in any year. That the publication of what was expected to be a high-impact manuscript was delayed to the second issue harmed this paper's citation record. If the principal investigator's suspicion is correct, for certain, misconduct has occurred. The same can be said of a reviewer who makes unreasonable demands of the authors for edits or additional work before publication can proceed. For sure, it is the peer reviewer's prerogative to give the opinion that more work needs to be done or serious editing needs to be done. However, if the reviewer is a serious competitor in the field and he or she is simultaneously working on a publication of his or her own, in all likelihood, he or she is acting with intentional malice and a conflict of interest. Fortunately, the editor can intervene and overrule such a review or even send out the manuscript to an additional reviewer if the review process is taking too long. Since the editor always knows who the reviewer is, a thoughtful editor can and will stop sending such a reviewer manuscripts to review. Unfortunately, however, some fields are too small to allow this to be done easily.

Peer review, however, is not a perfect system. It has its detractors as discussed and Web 2.0 may well be a solution. Peer review is also a potential "soft spot" that can be breached by misconduct. As an example, peer reviewers can be impersonated in some cases by an author of the very paper being reviewed by the author entering in a fake email address for a suggested reviewer. Such actions represent a very serious breach of the ethical code of all manners of research.

WRAPPING UP

Although there are clear drawbacks to the current peer-review system, it works and it usually works exceedingly well. It can be said that making rules or laws based upon rare occurrences is foolish. It is partially for this reason that the entire peer-review process has not be torn down and completely renovated. The main reason, again, is the presumption of trust. Those who abuse the system and behave unethically are few and far between, even though they get enormous levels of attention when they are caught. In fact, the fact that they *are* caught is perhaps the best proof that the system works. Therefore, even with the shortcomings of the current peer-review process, there is likely no alternative that would fix the current problems and not create a brand new set of problems.

4 What Constitutes Responsible Conduct from the Point of View of Human/Animal Subjects in Research?

In the mid-1900s, codes for the use of humans in research became more formalized, even if they weren't exactly widely shared and followed. The origin of the Nuremberg Code[1] is (believe it or not) *pre*-World War II Germany. Although it was not explicitly codified until the war crime trials after the war, the spirit of the Nuremberg Code was largely followed by German Physicians, including the principles of informed consent. Famously (or perhaps infamously) the Nazis either disregarded them or interpreted them as they saw fit to execute their deranged plans. After, or even as a partial result of the trials, this ten-point code came into existence:

1. Voluntary, well-informed, understanding consent of the human subject in full legal capacity is required.
2. Positive results for society that cannot be procured some other way should be the aim of the experiment.
3. Some manner of previous knowledge should be present to justify the experiment.
4. Unnecessary (which is not defined by the code) physical or mental suffering and injuries should be avoided.
5. If death or disabling injury is an implied risk, the research should not be carried out.
6. The humanitarian benefits should exceed the risks.
7. The facilities and other preparations/physical settings must adequately protect the subjects against the risks of the experiment.
8. Those conducting the research must be trained and scientifically qualified.
9. Any subject can quit at any time.
10. The medical staff overseeing the experiment must stop the experiment if they observe that continuing would be dangerous.

Although it is not in any way binding law and was also dismissed initially, much of this code still exists in more modern documents such as the Declaration of Helsinki

---

[1] https://en.wikipedia.org/wiki/Nuremberg_Code, last accessed 6/23/18.

(which most modern nations' human subjects research are based upon) and its forerunner, the Declaration of Geneva.[2]

The Declaration of Geneva as currently published by the World Medical Association states that a member of the medical community shall

- Dedicate his or her life to the service of humanity.
- Make the health and well-being of his or her patient his or her first consideration.
- Respect the autonomy and dignity of his or her patient.
- Maintain the utmost respect for human life.
- Not permit considerations of age, disease or disability, creed, ethnic origin, gender, nationality, political affiliation, race, sexual orientation, social standing, or any other factor that intervenes between his or her duty and his or her patient.
- Respect the secrets that are confided in him or her, even after the patient has died.
- Practice his or her profession with conscience and dignity and in accordance with good medical practice.
- Foster the honor and noble traditions of the medical profession.
- Give to his or her teachers, colleagues, and students the respect and gratitude that is his or her due.
- Share his or her medical knowledge for the benefit of the patient and the advancement of health care.
- Attend to his or her own health, well-being, and abilities to provide care of the highest standard.
- Not use his or her medical knowledge to violate human rights and civil liberties, even under threat.
- Make these promises solemnly, freely, and upon his or her honor.

It would not be hyperbole to consider this code of honor for medical professionals. The Declaration of Helsinki takes these to a far greater degree regarding human experimentation and is considered by many to be the cornerstone document on human research ethics.[3] Although not legally binding, its influence can be seen in many national or regional regulations governing human subjects research. In 1989, however, the United States stopped recognizing revisions and, in 2006, stopped referring to the declaration altogether. In 2008, the United States replaced it in its codes with "good clinical practice." The European Union also has been only citing the 1996 version in its clinical trials directive published in 2001, meanwhile the European Commission refers to the 2000 revision.

A large resource for those interested in the standards and regulations abroad regarding human research standards is the 2018 edition of the international compilation of human research standards, published by the Office for Human Research Protection of the U.S. Department of Health and Human Services. The extensive resource (which references an individual country's or international documentation,

---

[2] https://en.wikipedia.org/wiki/Declaration_of_Geneva, last accessed 6/23/18.
[3] https://en.wikipedia.org/wiki/Declaration_of_Helsinki, last accessed 6/23/18.

rather than summarizes regulations) is broken down by the following topics, for each country listed:

- General
- Drugs, biologics, and devices
- Clinical trials registry
- Research inquiry
- Social–behavioral research
- Privacy/data protection
- Human biological materials
- Genetic research

Genetic research then includes embryos, stem cells, and cloning. The FDA has outlined many rules, which can be found on the FDA's website. As regulations often change, and with them hyperlinks that lead to pertinent documentation, no links are provided here. The risk of the information becoming outdated or incorrect is too high.

Some of the data the FDA uses to evaluate the safety of any product to be used on humans or other animals are as follows:

- Toxicity
- Observed (demonstrated) lack of adverse side effects
- Risks of clinical studies with humans and other animals
- Any potential adverse effects, especially carcinogenic and teratogenic
- The level of use (dose and duration) that can be approved

Regarding human subjects testing, the seminal event (in the United States anyway) was the authoring of the Belmont Report,[4] created in the aftermath of the Tuskegee Syphilis Experiment, as an attempt to summarize the basic ethical principles behind the National Commission for the Protection of Human Subjects of Biomedical and Behavioral Research, created in 1974 with the signing into law of the National Research Act. This report made no attempt to make specific recommendations for administrative action. Instead, it attempted to provide a framework to resolve any ethical problems that manifest during research that involves human beings. It's important to note that these guidelines are operative whether it is drug-based research or even sociological or survey-based research. Institutions conducting research and receiving federal funding (of any type, including non-research funds such as financial aid) are expected to form committees to evaluate all research projects that will involve human subjects. It is the job of such committees to ensure that every possible safeguard for the participants has been taken. One of the outcomes of this was a basic set of ethical principles:

1. Respect for persons
   First, individuals must be treated as persons who can make decisions for themselves. Persons with disabilities that make them less autonomous

---

[4] www.hhs.gov/ohrp/regulations-and-policy/belmont-report/read-the-belmont-report/index.html, last accessed 6/23/18.

are entitled to some sort of protection. As a result, there are two moral requirements that must be met: acknowledge autonomy and protect those with lower autonomy.

Informed consent

Persons with autonomy must be given adequate information about the study such that they can make a determination regarding their *willing* participation in the study. It must also be clear that the volunteer understands the information that he or she was given, the risks associated, and the potential benefits of this study, both to themselves and for the greater good. An important point is that not only people of reduced mental facilities (i.e., Alzheimer's disease, Down syndrome) but also prisoners are considered persons with reduced autonomy. This is because prisoners may be more easily coerced into participation in a study they would not otherwise willingly participate in. Such individuals should receive extra protection to make sure they are not taken advantage of. One case of this was the use of unwilling prisoners in Nazi concentration camps during World War II. These test subjects unwillingly evaluated drugs that would then go on to help non-captives.

2. Beneficence

The Hippocratic maxim "do no harm" applies here as well as in medical ethics. Two rules have been thusly formulated to cover beneficence: (1) do not harm and (2) maximize possible benefits and minimize possible harms. The overseeing researchers must carefully evaluate the balance between scientific progress and an individual's suffering and well-being. Nearly all medical treatments have risks associated with them. Generally speaking, all of the associated risks are even more prevalent during the year(s) immediately after a treatment's deployment since the diversity of persons taking it expands more at this point than any clinical trial ever could match. In almost all the cases of medical treatment that we employ today, these risks were viewed to be minimal or rare enough that the benefits outweighed them. It should be made absolutely clear that a patient losing his or her life or becoming permanently damaged during the course of treatment or testing is not an indication of misconduct on the part of the researchers. This, however unfortunate, is part of scientific progress. If, on the other hand, obvious signs (or complaints) of suffering went ignored or evidence that suggested severe complications in animal testing studies (which always happen before human testing) was ignored, or misrepresented in reports to the FDA, misconduct on the part of the researchers is obvious. Misconduct is just as obvious on the FDA's part if the federal regulatory body is the one ignoring information. Naturally, if the FDA is bribed, both parties are behaving unethically.

3. Justice

Justice in this case asks the question "Who ought to receive the benefits of research and bear its burdens?" For this, a handful of widely accepted formulations have been agreed upon. These formulations are as follows: (1) to each person an equal share, (2) to each person according to individual need,

(3) to each person according to individual effort, (4) to each person according to societal contribution, and (5) to each person according to merit. For justice to be a part of the study, the study must not involve test subjects who are unlikely or unable to be among the beneficiaries of the therapy being investigated. In other words, persons in desolately poor areas cannot be used to test treatments they will be unable to afford themselves once they become available.

The report then goes on to talk about the application of the general principles to the conduct of research:

1. Informed consent
      Consent is a process broken down into the categories of receiving information, comprehending the information, and then volunteering for the study.
   a. Information
            Usually, the following must be made available to the subjects:
         i.   The research procedure
         ii. Their purposes, risks, and anticipated benefits
         iii. Alternative procedures (when therapy is involved)
         iv. Statement offering the subject the opportunity to ask questions and to withdraw from the research, even after the research has begun
                  There are of course studies that occasionally require that information about the research be withheld from the participants. Such is allowed if and only if it is clear that
         i.   Incomplete disclosure is truly necessary to fulfill the goals of the research.
         ii. There are no undisclosed risks that are more severe than minimal.
         iii. There is an adequate plan for debriefing subjects when appropriate and for dissemination of research results to them.
                  Under no circumstances should information about risks ever be withheld for the purposes of gaining the cooperation of the volunteer.
   b. Comprehension
            It is critical that the information be presented to the volunteers in a manner that will allow them to ask questions during the explanation and to fully understand what the study entails. For example, it would be completely inappropriate to quickly read 20 pages of material to the volunteer and not allow them the opportunity to read it themselves. In some cases, it would be appropriate to quiz the participants before beginning the study so that they can demonstrate a clear comprehension. In cases where somebody is of limited comprehension ability, a third party can stand in for them and make the decision. This third party should be able to understand the subject's situation and is able to also act in his or her best interest. Examples of individuals with limited comprehension are infants, young children, mentally disabled patients, the terminally ill, and the comatose.

c. Voluntariness

   Consent must be voluntarily given by the test subject. No threats or any other forms of coercion can be employed to gain the volunteer's cooperation with the study. Forms of coercion that would be inappropriate would be any form of penalty for not participating in the study and excessive rewards. As is the current standard for consent in sexual assault, consent to participate in such research can be revoked at any time, for any reason, and without explanation.

2. Assessment of risks and benefits

   One important aspect of the assessment of risks and benefits is that other modes of achieving the same ends must be put into the context of the present study.

   a. The nature and scope of risks and benefits and the systematic assessment of risks and benefits

      As mentioned above, the benefit-to-risk ratio must be in the favor of benefits. If the risks greatly outweigh the benefits, it would be improper to conduct the study, and the committee overseeing such work at an institution would not give the researcher approval. Certain single benefits may be viewed as more heavily weighted, especially in diseases where there is no viable alternative.

3. Selection of subjects

   With the exception of studies that seek to investigate a particular group, selection of the subjects cannot be made based upon social, racial, sexual, and cultural biases that persist in society. Furthermore, selection cannot be done on a basis that allows specific groups, only, to benefit. Careful oversight should be applied in such a case, either way. The Tuskegee Study, which involved syphilis-infected African-American men only, and even what the Nazis did to millions of Jews were both "research" (and **no**, I don't consider either—but especially not the latter—to be actual research) focused on one group. It is easy for such research to devolve into discrimination, racial profiling, or ethnic cleansing. That being said, sometimes, there is a gray area, and I can only encourage you to not do anything you're uncomfortable with. Let me be clear one more time: neither of the aforementioned cases are gray to me. Both are abhorrent abuses and unethical.

There are various issues, in addition to a strict adherence to the federal regulatory process outlined above that are worth discussing. First, the participants in the study must be afforded full disclosure of the hazards associated with the drugs during earlier testing. For example, if the earlier phase trials demonstrated that six of ten mice suffered strokes at high doses of the drug, the human test subjects **must** be told this. In fact, with such a negative safety profile, it would have to be *very* effective or offer some other significant benefits for the FDA to permit clinical trials of this drug, but that's beside the point here. They likewise must be told how efficacious the drug was in earlier animal testing. For certain, the participant is much more "science experiment" than patient, but full disclosure is imperative since it **will** influence someone's willingness to be a "science experiment." Undoubtedly, some people volunteer for such trials as a last chance at surviving the disease they are afflicted with, and recent Right to Try laws[5] in the United States may change that landscape. The more altruistic volunteers do so to help humankind. Undoubtedly, there are also some who do it for the love of science. Everyone is entitled to participate, or not, for his or her own reasons.

Also, everyone knows that some drugs have severe side effects. Some of these may involve varying degrees of pain or discomfort. Different people respond to pain differently, making pain something that is phenomenally difficult to measure. For example, somebody not accustomed to gastrointestinal illness may find a drug that gives him or her diarrhea for a week to be inhumane punishment. Furthermore, somebody may simply have a very low pain tolerance, making a drug that causes joint pain unbearable to him or her. This makes measuring the pain or suffering very difficult to approach ethically. That is too say: how much is too much, at what point would a participant be removed from the study to reduce his or her pain and suffering if he or she doesn't remove himself or herself, and how much pain must a tester experience before it must be reported as a "rare side effect," just to name a few important questions that must be answered. Also, some drugs are used despite very high toxicity. An example of this is the anti-HIV drug AZT, once one of the only treatments for this disease. AZT is known to have bone marrow toxicity. Sometimes, if a drug is one of the precious few that have efficacy against a death sentence such as HIV, pancreatic cancer, or inoperable brain cancer, even severe side effects may be tolerated to give the patient a chance. As long as the patient is not kept in the dark about these risks, nothing unethical has occurred. Cancer treatments, which commonly have led to hair loss, are also very well tolerated because such a side effect is considered small when compared to the benefits of the treatment survival. Also, when (and truly not if) safer drugs are invented, they supplant those with less favorable safety profiles.

Regarding animal research, pain and some other side effects (i.e., lucid dreams) become nearly impossible to measure accurately and this is why some side effects are not found before the drug is given to people in addition to animals not *being human*. For certain, most people are observant and aware enough to be able to look at an animal and know it is in discomfort. What is not always immediately clear, however, is what's wrong. Something like a limp is obvious enough, as is diarrhea, but a dog cannot say "I have a headache," a cat cannot say "my throat hurts," a horse cannot say "my stomach is killing me," hamsters don't lament "you would not believe the dream I had," and the list can go on, quite easily. Therefore, such negative side effects, and especially their severity, become almost impossible to measure during animal testing making the humans "guinea pigs." And yes, I both see and mean that pun.

As discussed earlier, the United States Department of Agriculture does not dictate research protocols; it forces the facilities to establish its own ethics oversight committees that guide its actions and decisions governed by the

---

[5] https://en.wikipedia.org/wiki/Right-to-try_law, last accessed 6/23/18.

Animal Welfare Act. The Animal Welfare Act blue book[6] sets the following guidelines:

1. Adequate care and treatment for housing, handling, sanitation, nutrition water veterinary care, and extreme conditions must be provided by the facilities.
2. Dogs must be provided opportunities to exercise.
3. Primates must be provided opportunity for psychological well-being.
4. Anesthesia or pain-relieving medication must be provided to minimize pain or distress.
5. Unnecessary duplication of specific experiments using regulated animals is prohibited.
6. The establishment of an institutional animal care and use committee that will oversee the use of animals in the experiments.
   a. This committee is then responsible for ensuring the facility complies with the Animal Welfare Act and providing documentation of compliance with the Animal and Plant Health Inspection Service.
   b. The committee must contain at least three members and membership must include one veterinarian and one person not affiliated with the facility.

Another important question about animal testing is "which animals count?" The reality is that some animals count more than others. For example, cold-blooded animals are exempt from coverage. Specifically, vertebrates count more than invertebrates, and mice are near the bottom of the totem pole of vertebrates. Vertebrate animals (even mice) require vigorous prior approval and careful monitoring, whereas invertebrates such as insects require no approvals or monitoring. Some people, without doubt, will have a moral problem with animal testing and this is their right. A former coworker of mine once (seriously) stated that animal testing should never be done, preferring to just test things on people. The fact of the matter remains that there are well-established and agreed upon rules that, when followed, avoid a form of scientific misconduct that is not scientific misconduct in the same way that was discussed earlier (though its similarity to human subjects research should go without saying).

These two separate documents[7] govern the care and use of animals in terms of both research and teaching. As these documents specifically cover researcher practices and/or teaching, they do not cover legalities such as animal cruelty. These sorts of regulations are covered by regional or even national law. The document containing the regulations for agricultural animals in research and education, referred to as the Ag Guide, contains the following chapters and topics:

Although an in-depth description of all these regulations is well beyond the scope of the book, a brief description of what some of the terms mean or entail is appropriate.

---

[6] www.aphis.usda.gov/animal_welfare/downloads/AC_BlueBook_AWA_FINAL_2017_508comp.pdf, last accessed 6/23/18.
[7] Guide for the care and use of laboratory animals, 8th edition, national research council of the National Academies, 2011. Guide for the care and use of agricultural animals in research and teaching, 3rd edition, Federation of Animal Science Societies, 2010 (the Ag guide).

Chapter 1: Institutional policies	Chapter 2: Agricultural animal health care	Chapter 3: Husbandry, housing, and biosecurity
Monitoring the care and use of agricultural animals	Animal procurement	Facilities and environment
Protocol review	Veterinary care	Feed and water
Written operating procedures	Surgery	Husbandry
Animal health care	Zoonoses	Standard agricultural practices
Biosecurity	Residue avoidance	Handling and transport
Personnel qualifications	Restraint	Special considerations
Occupational health	Transgeneic and genetically engineered and cloned animals	Biosecurity
Special considerations	Euthanasia	Biocontainment
Chapter 4: Environmental enrichment	**Chapter 5: Animal handling and transport**	**Chapters 6–11: Beef cattle, dairy cattle, horses, poultry, sheep and goats, and swine**
Cattle	Biomedical versus agricultural research requirements	Facilities and environment
Horses	Flight zone and behavior principles	Feed and water
Poultry	Aids for moving animals	Husbandry
Physical enrichment	General principles of restraint and handling principles to prevent behavior agitation during restraint for all species	Standard agricultural practices
Sheep and goats		Environment enrichment
Swine		Handling and transport
General considerations	Recommendations for all species transport	Special considerations (absent in horses)
		Euthanasia

PERSONNEL

Similar to human subjects research, the personnel conducting research in which they handle nonhuman animals must be qualified. Training should address the following:

1. Husbandry needs, proper handling, and surgical (pre, post) and perioperative procedures
2. Methods for minimizing stress and pain in animals and the number of animals
3. Methods for reporting deficiencies in core program
4. Use of information services
5. Methods of euthanasia

ENVIRONMENTAL ENRICHMENT

This involves the enhancement of animals' physical or social environment and also involves social, occupational, nutritional, physical, and sensory enrichment.

HOUSING

There are a number of considerations when determining the housing of animals. Obviously, the larger the animal, the larger the space needed. Also, temperature, ventilation, and water vapor pressure must be considered. Excreta, access to feed and water, and air quality all further contribute to this small sampling of the regulations.

BIOSECURITY

Historically, in feed animals, this was defined as security measures taken to prevent the unintended transfer of pathogenic organisms and subsequent infection by humans, vermin, or other means. Agricultural research has adopted these and, with the concerns about bioterrorism, added protection against intentional contamination.

BIOCONTAINMENT

If pathogens involved in the research are classified as a pathogen "of veterinary significance" in the Centers for Disease Control and Prevention (CDC) book *Biosafety in Microbiology and Biomedical Laboratories*, the facilities must meet specific criteria for their design, operation, and containment features.

In the general guide for the use of laboratory animals, the "three Rs" can be found: replacement, which refers to methods that avoid animals; refinement, which focuses on animal well-being and comfort; and reduction, which involves strategies for using as few animals as possible in the research.

Additionally, this guide covers the following chapters and topics in its table of contents. As before, a wider discussion of these is beyond the scope of this book.

Chapter 1: Key concepts	Chapter 2: Animal care and use program	Chapter 3: Environment, housing, and management
Applicability and goals	Regulations, policies, and	Terrestrial animals
Intended audiences and uses of	principles	Terrestrial environment
the guide	Program management	Microenvironment and
Ethics and animal use	Program management	macroenvironment
The three Rs	responsibility	Temperature and humidity
Key terms used in the guide	The institutional official	Ventilation and air quality
Humane care	The attending veterinarian	Illumination
Animal care and use program	The institutional animal care	Noise and vibration
Engineering, performance, and	and use committee	Terrestrial housing
practice standards	Collaborations	Microenvironment (primary enclosure)
Policies, principles, and	Personnel management	Environmental enrichment
procedures	Training and education	Sheltered or outdoor housing
Must, should, and may	Occupational health and safety	Naturalistic environments
	of personnel	Space
	Personnel security	Terrestrial management
	Investigating and reporting	Behavioral and social management
	animal welfare concerns	Husbandry
	Program oversight	Population management
	The role of the IACUC	Aquatic animals
	IACUC constitution and function	Aquatic environment
	Protocol review	Microenvironment and
	Special considerations for	macroenvironment
	IACUC review	Water quality
	Postapproval monitoring	Life support system
	Disaster planning and	Temperature, humidity, and ventilation
	Emergency preparedness	Illumination

(Continued)

In a related topic to drug testing, in some cases, a close examination of some drug development literature will reveal something that the lay reader may interpret as grotesquely unethical. Sometimes, the drug that is pursued is **not** the one with the greatest *in vitro* activity and even not the highest *in vivo*[8] efficacy. There are several perfectly good science and good ethics reasons why this would be the case. First, a closer reading sometimes will reveal that the more active drug has significant toxicities associated with it as well. This is certainly possible and very unfortunate. In this sort of case, it is the only viable option to avoid this more active drug. Second, there will also be cases where a less active drug will ironically be active against a wider range of strains of a disease. As an example, a drug may have a record level of activity against a single strain of HIV, but minimal activity against many others. Compare this to a less active drug, one that is roughly equiactive against all strains of HIV. In this case, the "less active" drug would logically be the one pursued since it would have the broadest effectiveness. Finally, this last instance is *not* immediately apparent in the paper, and there may be synthetic difficulties that dramatically increase the costs of drug production. These difficulties may range from low-yielding reactions to costly purifications or dangerous synthetic protocols or intermediates. You may be tempted to say, "so what", since after all, it is the company's job to produce these drugs. Although this is not necessarily incorrect, nor is it unfair, *per se*, it is the company's prerogative to make a profit. The more money it costs to produce large quantities of the drug, the more money the treatment will cost. With this in mind, synthetic difficulties that triple the cost of production are certainly viable reasons to avoid a drug candidate as they would drastically increase the cost of the drug. When one recognizes that some diseases, for example, malaria and HIV, are most devastating to people who are poor beyond comprehension, the need for low-cost medication (even with the assistance of philanthropists like Bill and Melinda Gates) is obvious.

An other case is when there are two candidates, we'll call them drug A and drug B, which are being pursued as anti-HIV treatments. Let's assume that drug B is five times more active than drug A. Let's also assume that drug A (the less active one) is 15% more active than every other drug on the market but one, and this one has complications that neither drug A nor drug B suffers from. In this sort of hypothetical case, drug A may (and perhaps *should*) be released first. The reason is that after resistance to drug A is acquired by the virus (which in HIV's specific case is almost inevitable), drug B can then (at least potentially) be deployed to replace it. This sort of tactic *usually* does not work in the reverse direction. Similar cases are seen with methicillin-resistant staphylococcus aureus (MRSA). Cases of severe bacterial infections are increasing in the public eye. There are, in fact, some antibiotics that do combat some of these infections. When an infection is present in a patient, the antibiotics are used in a specific order. **This is not unethical**; it is simply smart science and medicine. Immediately deploying the antibiotics that represent the last line of defense will only breed strains resistant to these antibiotics as well, resulting in a complete catastrophe. In any case, all the data from the drug candidates must always be presented in especially patent applications, even if the candidate will not be used.

---

[8] *In vitro* refers to assay studies done on cell cultures, while *in vivo* refers to studies done in living organisms.

WRAPPING UP

Whatever one's feeling is toward drug testing, whether it is human or animal testing, it is necessary. Without testing, we would have no ability to identify which drugs work and which ones don't. For this reason, the FDA and its counterparts in other countries have very rigorous regulations that govern the process. When the system is followed, it works very well and minimizes the suffering that may be associated with the testing of drugs. Furthermore, drugs must be continually monitored for safety, even after deployment, and we must take care to not lose some of our best hopes for recovery to disease resistance.

5 The Ethics of the Pharmaceutical Industry

The pharmaceutical industry, or Big Pharma, as it is often called, is one of the most critically important industries in our society. With the reputation that Big Pharma has for being greedy and rich (even the name Big Pharma is not doubt an attempt by its detractors to elicit a David vs. Goliath image), many may be surprised to hear that there is not even one pharmaceutical company in the top ten on the Fortune 500 list.[1] In fact, there is not even a single pharmaceutical company in the top 30! The first to weigh in is Johnson & Johnson at #35. Pfizer is next at #54, Merck at #69, and Gilead Sciences at #92, and the only others in the top 100. Giants Eli Lilly and Bristol-Myers Squibb barely make the top 150 at #132 and #147, respectively. Thus, I encourage you to not believe everything you read and hear about this industry. For sure, they make plenty of mistakes; they are after all run by people. But this book is not about defending the pharmaceutical industry. I'll leave that to others and I'll do the same regarding attacking it. Herein, some of the basic rules governing how this essential industry develops new drugs will be covered.

Before moving on to the science ethics, it is relevant to talk about some of the financial aspects of this industry. Although the number is disputed, the cost to develop a drug (i.e., from the concept to the market) is around $2 billion.[2] Recovering these costs is part of the price of the treatments, naturally, since without recovering these costs, the companies could not stay in business. Although we can certainly argue whether or not forgone profits should be counted in the cost of drugs while they are in development (they currently are), what may not be immediately clear to a nonscientist or someone who is generally lacking in his or her scientific literacy is that this cost does (and this should be without argument) include all the failed analogs, the testing and preparation of which also costs huge sums of money. Often, these analogs never come close to clinical trials before they fail and those that fail in clinical trials are that much more costly. After the pharmaceutical company sets its price, pharmacies such as CVS and Rite Aid, which **are not pharmaceutical companies**, add their markup so that they could also stay in business. Then, insurance companies decide not only how much but *what* they will pay for while the consumer pays whatever is left, usually as a nominal copay though sometimes much more. Although some treatments carry a very large price tag set by the pharmaceutical company, they are not the only factor in the price to the consumer. Pharma is but one variable in the equation.

---

[1] http://fortune.com/fortune500/2017/list/filtered?industry=Pharmaceuticals.
[2] Forbes.com, The Cost of Developing Drugs Is Insane. That Paper That Says Otherwise Is Insanely Bad, by Mathew Herper, October 16, 2017.

Drugs that are for rare diseases oftentimes carry a far larger price than those that are more common. This is because the cost of developing that drug is not proportionally lower due to its smaller patient pool. Since the cost of developing these drugs is still in the hundreds of millions, if not in excess of $1 billion, the pharmaceutical companies must balance their sheets somehow. With a smaller patient pool purchasing the drug, a higher price is needed to do this. While, on the one hand, the insurance companies could just pay for these treatments, on the other, they in turn would pass that cost down to the consumers, and it would likely be in the form of higher premiums for everyone, rather than just the persons using the expensive treatment. You see, it is a vicious cycle.

The philanthropy of the pharmaceutical industry also often goes unrecognized by the general public. Perhaps part of why it goes unrecognized is that they do much of this work in third-world countries, rather than locally. For example, Pfizer donated more than 500 million doses of a medicine to treat trachoma, a terrible eye disease that causes irreversible blindness from the damage caused by the infected persons' eyelids turning in and scraping the eyeball.[3] Johnson & Johnson donated enough of its deworming pill mebendazole to treat children with intestinal worms in eight countries while also partnering with governments and organizations to stop reinfection through hygiene and improved sanitation and clean water.[4] These are just a small sample, but none of this excuses any of the missteps of the industry, naturally.

But what are the rules for the pharmaceutical industry? There are of course many aspects of this giant and diverse industry to consider. In all likelihood, this topic could fill an entire text. To fit it into one chapter, some decisions must be made and limits must be set. This book limits the discussion to the following topics:

- The procedure for getting a drug approved
- Direct to consumer advertising

THE APPROVAL OF A DRUG[5]

Assuming that the drug has already been optimized for desired activity and proven to be safe (enough) in animals during preclinical studies, Phase I clinical trials are generally when a compound is first administered to humans. During this part of the process, the drug is given to tens of healthy (fully informed) volunteers. It is unethical to give people medicine they don't need, you say? Why healthy volunteers, you ask? Taken in turn, NO, and because this is the best way to identify the side effects of the drug. Do understand that these volunteers know the risks. They would be told of any side effects observed during animal testing and would need to be told if preexisting drugs with the same mechanism of action have any known side effects. The healthy qualification is because diseased persons may owe the observed side effect to the disease, rather than the medicine. This would yield a false positive for the drug

[3] www.pfizer.com/responsibility/global_health/international_trachoma_initiative, last checked 6/23/18.
[4] www.jnj.com/caring/patient-stories/keeping-children-free-of-infection, last checked 6/23/18.
[5] The FDA's Review Process: Ensuring Drugs are Safe and Effective, www.fda.gov/drugs/resourcesforyou/consumers/ucm143534.htm, last checked 6/23/18.

candidate and may ultimately prevent an otherwise viable medicine from reaching the market. Phase I, therefore, is where the safety of a drug is ascertained. It is also where excretion (urine or feces), metabolism, and frequency of dosing or other pharmacokinetic or pharmacodynamic information is first collected. If the drug clears this hurdle, it is then administered to fully informed diseased volunteers. In this second phase of clinical trials, the efficacy of the drug is determined for the first time. These trials usually break into the few hundred patient level. Safety is also monitored in this phase and continuously. Phase III is similar to Phase II, but it is significantly larger. The FDA and the drug sponsor will collaborate on how large the study must be. Phase II is a "soft rollout" of the drug, while Phase III is when they expand the pool of patients to as closely as possible approximate the eventual patient pool. Phase III would only be advanced to if Phase II testing demonstrates effectiveness at treating the disease. With this broader administration, not only is its efficacy measurement more reliable, but also the safety (which is monitored from the moment it is administered to any animal through postmarket monitoring) measurements are more reliable and informative. To briefly explain how it does so, consider that Phase II clinical trials only contain 100–300 people, while Phase III trials can reach into the thousands. The diversity allowed by the larger pool that can be attained during Phase III trials alone will give a better picture of both the efficacy and safety of a drug. During Phase II trials, it is possible, but during Phase III, it is likely that patients with multiple conditions and/or taking multiple drugs will be enrolled in the study. Also, a wider variety of ethnic backgrounds will be exposed to the medicine, and there are some traits that are more prevalent to some ethnicities than to others, leading to different responses to medicines and disease. This gives information about the interaction between drugs and is a more accurate simulation of how the drug will be used in society. Once it clears these hurdles and reams of paperwork are completed, the FDA considers all the data and renders a decision about the drug. If it is approved, it is allowed to enter the market and be used.

Curiously, doctors are permitted to prescribe an approved medicine for any purpose they think that it will work, even purposes other than those which a drug is approved for. The manufacturer is not allowed to advertise any such uses, and insurance companies may not pay for it. Also, ethical doctors do not do this wantonly. They only do so if there is good reason or solid evidence that it will work. This evidence may be in the form of published research by other doctors using the drug for this off-label use or a hunch the doctor has, based on his or her knowledge of how the medicine works.

The pharmaceutical industry has also come under fire for some of the perceptions around clinical trials: some of them are more observation than perception, in fact. It should be known that it is the pharmaceutical company that runs the clinical trials. It also pays for them entirely. The company pays the volunteers taking the medicines (especially in Phase I) and all the medical staff who administer the drugs and evaluate their efficacy (i.e., who study the patients). All the data are reviewed by a review board and ultimately by the FDA. Some find it unethical that the company has such a hand in the clinical trials. However, there is virtually no other option. Any other model would either cost the taxpayers enormous sums of money if the FDA or some other government agency were to run the trials or be extremely inefficient. Others point to how since the company is in control of the data sent to the FDA, they could easily obfuscate or even falsify results that would jeopardize the drug's entry into the market. There are hefty fines if companies do this, not to mention lawsuits.

Other criticisms come from who is enrolled in testing, in particular, the inclusion of women and children in clinical trials. The involvement of women and children in clinical studies has been reviewed, for example, the work by Joseph[6] and Liu.[7] Both discuss the need for each respective group to be better represented in clinical trials. This is on the basis of the fact that women and men have different biologies due to the differences between sexes and adults have different biologies than (especially pre-pubescent) children, including different metabolic rates. Historically, these groups are less represented in clinical studies due to ethical concerns. For example, children are one of the protected groups regarding informed consent since they typically lack the maturity and intellect to understand the risks. This would be especially true in Phase I trials. Women, of course, can get pregnant (within certain time frames of their lives), while men lack this ability. This understandably brings about additional ethical concern for several reasons. For example, if a woman becomes pregnant during the clinical study, it may be advisable for her to resign from the study to protect her gestating child. She may even be forced to; her pregnancy may invalidate her eligibility for the clinical trial for various scientific or ethical reasons. If this occurs or is discovered later in this person's involvement in the clinical study, it may be very costly. Also, a drug may damage a woman's egg cells, setting the stage for birth defects for a child conceived later in her life. These risks are viewed as being very great. On the other hand, one can logically ask the question, "Do we want to find all these problems before a drug gets on the market or after?" A logical way forward may very well be to include women in trials where animal studies showed no signs of birth defects but proceed more carefully with those that do. What this greater care may look like, I cannot propose. Greater inclusion of children is equally necessary but harder to do so safely.

CATEGORIES OF DRUGS

The United States Food and Drug Administration attaches a number of different attributes to drugs that are also worth mentioning here. It turns out that some drugs earn the company developing some of them select benefits.

ORPHAN DRUGS AND RARE DISEASE

The rare disease[8] program exists to facilitate, support, and accelerate the development of treatments for persons with rare disorders. How a rare disease is defined is different in some countries, but it should not be difficult to understand what it is; it is a disease that afflicts relatively small numbers of persons. The Orphan Drug Act of 1983 defined a rare disease in the United States as the one that affects less than

---

[6] Joseph, P. D.; Craig, J. C.; Caldwell, P. H. Y. *Br. J. Clin. Pharmacol.*, **2013**, 357–369.
[7] Liu, K. A.; DiPietro, Magen, N. A. *Pharm. Pract.*, **2016**, 708–716.
[8] www.fda.gov/forindustry/developingproductsforrarediseasesconditions/default.htm, last accessed 6/23/18.

200,000 persons in the country. It also provided tax breaks for companies pursuing such drugs since their product would have little to no potential for making enough money to justify its development and subsequent production, since these medicines are nevertheless needed.

NEGLECTED TROPICAL DISEASE

Neglected tropical diseases[9] are most prevalent in the developing world, although some cases do occur in the more affluent countries as well. These diseases are fundamentally different from rare diseases. According to the World Health Organization, the following are neglected tropical diseases:

- Buruli ulcer
- Chagas disease
- Dengue and chikungunya
- Dracunculiasis (guinea-worm disease)
- Echinococcosis
- Foodborne trematodiases
- Human African trypanosomiasis (sleeping sickness)
- Leishmaniasis
- Leprosy (Hansen's disease)
- Lymphatic filariasis
- Mycetoma, chromoblastomycosis, and other deep mycoses
- Onchocerciasis (river blindness)
- Rabies
- Scabies and other ectoparasites
- Schistosomiasis
- Soil-transmitted helminthiases
- Snakebite envenoming
- Taeniasis/cysticercosis
- Trachoma
- Yaws (endemic treponematoses)

Similar to rare diseases and orphan drugs, incentives are offered to companies to produce these drugs. Among the incentives available, a voucher can be acquired by the drug developer to obtain a priority review for any drug application he or she chooses. Priority reviews, covered briefly below, are in place to try to speed up the drug approval process and get important drugs to the market more quickly.

FAST TRACK

This process[10] is designed to facilitate the development and expedite the review of a new drug that addresses a serious condition and fills an unmet medical need. Although what constitutes a serious condition is a judgment call, if the drug is going

---

[9] www.who.int/neglected_diseases/diseases/en/, last accessed 6/23/18.
[10] www.fda.gov/ForPatients/Approvals/Fast/ucm405399.htm, last accessed 6/23/18.

to have a major impact on survival and day-to-day functioning of the patient, or that if left unchecked, the condition may become more serious is enough to be considered a serious condition. Obvious diseases are HIV/AIDS, Alzheimer's, heart failure, and cancer though other diseases such as epilepsy, depression and diabetes are also placed into this category. If there is no therapy that currently exists or if the new therapy has the potential to perform better than the available therapy, it would fall into the category of filling an unmet medical need. Some of the advantages that the new drug can show over available therapies are as follows:

- Superior effectiveness
- Less serious side effects
- Improved diagnosis
- Decrease in clinically significant toxicity that often causes treatment to stop
- Addressing an emerging or anticipated public health need

The benefits of a Fast Track designation are as follows:

- More frequent meetings with the FDA during the process toward approval
- Increased feedback from the FDA about the clinical trial design and use of biomarkers
- Eligibility for accelerated approval and priority review, if criteria are met
- A rolling review that allows sections of the application to be submitted as completed

A drug candidate can be placed on the Fast Track at the request of the company that created the drug. This request can come at any point during the process, and the FDA will decide within 2 months. The frequent communication, though it may seem like an added burden, ensures that questions and concerns are addressed more quickly and thus gets the drug to the market faster, assuming, of course, that the drug is worthy of being granted entry into the market.

BREAKTHROUGH THERAPY

A Breakthrough Therapy designation[11] is conferred on a drug if it is intended to treat a serious condition, and there is preliminary clinical evidence that suggests that the drug may demonstrate a significant improvement over available therapy. The magnitude and duration of the curative event, along with the importance of the clinical outcome, are considered when determining if the improvement is significant. A clear advantage over available therapy must also be shown in the preliminary clinical evidence. The clinically relevant endpoint can refer to an endpoint

- That measures the effect on irreversible morbidity or mortality.
- That measures the effect on symptoms that represent serious consequences of the disease.

---

[11] www.fda.gov/forpatients/approvals/fast/ucm405397.htm.

Other clinically significant endpoints may refer to findings that suggest that serious symptoms or irreversible morbidity or mortality is affected, including the following:

- An effect on an established surrogate endpoint
- An effect on a surrogate endpoint or intermediate clinical endpoint that will probably predict a clinical benefit
- An effect on a pharmacodynamic biomarker that isn't an acceptable surrogate endpoint but does suggest the potential for a meaningful effect on the disease
- A better safety profile

Drugs that are designated as Breakthrough Therapies gain the following benefits:

- Fast Track designation features
- Greater collaboration with the FDA as early as Phase I
- Organizational commitment involving senior managers

Like the Fast Track designation, Breakthrough Therapy designation must be requested by the company for it to be granted. However, the FDA can suggest to the company that it requests the Breakthrough Therapy status if the FDA feels that the submitted data suggest the drug meets the criteria and that the program would benefit (as would society) from the designation.

PRIORITY REVIEW

Under the 1992 Prescription Drug User Act, the FDA established a two-tiered system for review times of new drug applications: standard review and priority review.[12] This was done to improve the drug review process. While a standard review typically takes about 10 months, a priority review is completed in 6 months. The FDA decides the review designation, but an applicant can expressly request it. The clinical trial period is not shortened if priority review status is conferred. Such designation is reserved for cases where significant improvements in the safety or effectiveness of the treatment, diagnosis, or prevention of serious conditions are expected. Significant improves may include the following:

- Evidence of more effective treatment, prevention, or diagnosis of the treated condition
- Elimination or substantial reduction of side effects that otherwise limit or reduce treatment
- Better patient compliance with the treatment
- Expansion of the subpopulation for which there is safety and effectiveness evidence

---

[12] www.fda.gov/ForPatients/approvals/fast/ucm405405.htm, last accessed 6/23/18.

ACCELERATED APPROVAL

Accelerated approval[13] is different from priority review. Priority review basically sends the drug to the front of the line, while accelerated approval truncates the clinical process at least slightly. The drug development process can sometimes take as long as a decade or longer. Although this includes the preclinical work, the clinical studies can take years to complete. Also, in some instances, it may not be possible to arrive at a *bona fide* clinical endpoint. Cholesterol-lowering and other preventative medicines are very good examples of this since they are often taken for many years by the patient and ideally prevent a particular ailment. Continuing to use cholesterol-lowering medicines as the example, these drugs reduce the incidence of heart attacks and strokes by lowering the lipid profile of the patient. Since high lipid profiles are common in individuals who suffer these ailments, healthier lipid profiles (which these medicines help patients to achieve) serve as a surrogate endpoint for studies of such medicines. Otherwise, the clinical studies to determine effectiveness would last for decades, potentially. In 1992, the FDA-approved regulations that allowed drugs for serious conditions that filled an unmet medical need to be approved based on a surrogate endpoint. Although this certainly allows for the approval of drugs to be conferred more quickly, they do come with their risks. As life spans extend, patients taking these drugs will be potentially taking them for longer periods of time than any clinical trial would ever run, even sans accelerated approval. As a result, the risk of side effects is high, but there is nothing that can truly be done to resolve this. This is, in fact, one of the risks associated with such medicines. There is virtually no way to ensure that it can be safely taken for several years, continuously, until large numbers of people do so. The benefits of such drugs reaching the market outweigh the risks, however, in the opinion of the regulators. Finally, if after approval the drug fails to show the anticipated benefits, the FDA can rescind approval.

DIRECT-TO-CONSUMER ADVERTISING

Currently, only the following four countries permit direct-to-consumer advertising[14]:

- New Zealand
- The United States
- Brazil
- Hong Kong

In the United States, drugs can be advertised directly to the consumers for only the purpose(s) the drug has been approved, even if a doctor can prescribe it for other purposes too. Drug companies must also include some information, at least, about the side effects of the treatment. There is a wide range of arguments that can be made both in favor and against this form of advertisement. A full consideration of these is beyond the scope of this book, but a small set can be considered.

[13] www.fda.gov/forpatients/approvals/fast/ucm405447.htm.
[14] https://en.wikipedia.org/wiki/Direct-to-consumer_advertising.

Taking an argument that it allows the public to be more informed of their choices first, it is certainly true that if the pharmaceutical company can "speak" directly to the potential patient/consumer, there is no filter that modulates the information they receive. However, a counterargument can easily be made that such ads are being delivered to a public that may not fully understand what they are seeing and listening to. That is, they can easily be deceived and duped into wanting a drug.

Another argument is that it may inspire a patient to ask his or her doctor about a specific treatment for symptoms he or she may be experiencing but didn't know there were treatments for. Ignoring a moment that this is emblematic of another, entirely different problem in the doctor–patient relationship, this is a good thing. *Anything that gets patients into their doctors' offices and gains them access to potentially needed treatments is a good thing.* The claims that this leads to an overprescription of drugs do not hold water since, although doctors give away lollipops, they do not just give away prescriptions without a medical indication being present and those who do are not practicing good medicine. It is *their* fault, not the fault of the pharmaceutical company. A complication does come, however, from this. Patients may start to psychosomatically see the symptoms. Not only that but patients who go into their doctor's office to request a specific treatment are doing themselves a grand disservice and perhaps even harm if they intend to abuse the drug. A patient is far better served to go to see he or she doctor about his or her *symptoms*, not about the drug he or she think will help alleviate his or her symptoms or whatever condition is causing the symptoms. That being said, a good doctor will not fall for this and diagnose the patient and prescribe the drug most suited for his or her diagnoses. Anything other than that is the doctor's fault, not the pharmaceutical company and certainly not the patient. To be fair to doctors everywhere, in localities where the patient-to-doctor ratio overwhelms a doctor, it may not be he or she "fault" to fall into a pattern that sees as many patients as possible in as a little time as possible. This, however, is indicative of an entirely different but no less urgent sociological problem.

With the concerns surrounding opioid addiction, the pharmaceutical industry is under pressure to also play a role in limiting the abuse of these necessary medicines. In fact, the FDA has taken the stance of rejecting certain drugs that run a risk (in its view) for abuse or overdose. Such was the case with the drug Moxduo,[15] a combination of oxycodone and morphine. One of the tactics that pharmaceutical companies can employ to ease fears is to incorporate abuse-deterrent properties.[16] A number of tactics are possible. One is to create barriers to prevent the extraction of the active ingredient or to make the pill/capsule harder to crush so that the drug cannot be inhaled. Other tactics include adding something otherwise non-active that makes the abuser uncomfortable with the route of use. All these tactics, without doubt, make the public as a whole safer, but they also all increase the price of the treatment. Nevertheless, it is one of the ways that the pharmaceutical company can ensure that its products are safe. It is part of its obligation to the public in exchange for the trust conferred in it by a public who quite literally cannot live without it.

---

[15] www.apnews.com/94c15d0009594469a5244bafcd078520, last accessed 6/23/18.

[16] www.pharmacytimes.com/publications/issue/2017/august2017/abusedeterrent-formulations-one-tool-in-the-opioidabuse-epidemic.

WRAPPING UP

The pharmaceutical industry, without doubt, has its flaws. The federal government, particularly the FDA, oversees the marketing of this industry's products. There are strict guidelines and many safeguards built into this process to keep us, the consumer-patients, safe.

6 Science and the Public

On the one hand, in the case of publicly funded research (research that is funded by federal or state tax dollars in the U.S., anyway), it can easily be argued that the researchers have an obligation to show the taxpayers what it is that they have paid for. It is exactly this argument that has led to a repository of manuscripts from federally funded research, even those found in subscription journals. On the other hand, the public is comprised largely of nonscientists who simply are not equipped to understand and evaluate the research being reported. Consequently, in all instances, it is better to put the research through the test of peer review before "going public" with results. This allows for some level of quality control of science **by scientists**. In matters of extreme emergencies, the peer-review process can be sped up, but the limiting step will always be the speed of the science. Though unrelated to scientific misconduct, one thing that is for certain is that scientists and not the lay public **must** remain in control of funding decisions. The lay public simply does not have the experience, the education needed to determine the quality, or the potential impact of science. Therefore, they and their representatives in government must not be part of the funding decisions.

MEASLES, MUMPS, RUBELLA, AND AUTISM

One example of disastrous consequences of scientific misconduct involves the "link" between autism and vaccines. This case is discussed in more detail in Chapter 10, but a discussion about its impact on society is relevant here. Although the study that initially demonstrated a link between the measles, mumps, and rubella (MMR) vaccine and autism has been retracted from the journal *Lancet* amid the discovery of fabricated data and other misconducts, it has had far-reaching impacts on society. It remains in the public psyche to some extent that the MMR vaccine causes autism and that vaccines, in general, are unsafe. People may be surprised today to hear that no such link between MMR and autism is confirmed to exist. There are no credible research studies that support such a link. The repercussions of this now discredited study have been felt for years and will likely be felt for many more. Parents have been frightened into not vaccinating their children out of a fear that it may cause autism in their children. This has the result of infectious diseases coming back, infecting, and killing children who (in all likelihood, anyway) would have been vaccinated if this violation of public trust had not occurred. Widespread vaccination also imposes a herd immunity effect, important to keep individuals unable to receive vaccines safe. Thus, although this is not necessarily a case where researchers went public too soon, it perfectly demonstrates the awesome power to affect society that science possesses. This power must be handled with extreme care. There is a wonderful series of articles by Brian Deer[1]

---

[1] *British Medical Journal (BMJ)*, **2011**, *342*, C7001. *BMJ*, **2011**, *342*, 5258. *BMJ*, **2011**, *342*, 5347.

that very effectively, clearly, and fairly document this case. It (the case) is filled with bad science, fabricated data, omitted data, and alleged conflicts of interest. The reader interested in reading about a case that has a variety of infractions, along with the penalties to the researchers and society, should read Deer's articles.

CLIMATEGATE

It is impossible to not be aware of the controversies that have surrounded climate change. For the purposes of this discussion, whether climate change is real is irrelevant. What is relevant, however, is that the exposure of what has come to be known as *Climategate* has severely damaged the credibility of the scientists investigating and advocating climate change in some of the public eye. Years later, the integrity of the researchers and the soundness of their conclusions are widely and openly questioned by factions of the public. As a result, even convincing evidence of the reality of climate change is less likely to be accepted by some parts of the public and acted upon. When the importance of the science is as high as it is with global climate, that is, it affects the entire planet, trust in those who are performing and presenting the work is of the utmost importance. This violation of trust in scientists is an extreme case that is presented to demonstrate the disastrous consequences that could arise from scientific misconduct. That the important policy decisions about this (and other topics) are usually made by one of two extreme sides of the argument only further puts frankly all humanity at risk. One can easily argue, however, that any side that ignores such risks for economic (rather than scientific) reasons is behaving unethically though not in the scientific misconduct sense.

HIV VACCINE

In a less damaging case, in September 2009, the first (partially) successful HIV vaccine was reported. This report was made a few weeks to a month before the study was published in the *New England Journal of Medicine*. The timing of the events suggests that the paper was already through the peer-review process at the time of the original public announcement. The researchers in the public announcement did not initially release the results of all the trials done in the study. It turns out, and this was revealed after the journal published the study and the results were presented at a conference, that there were two additional analyses that argued less strongly in favor of a successful vaccine. Let's be clear on a few things:

1. No data was fabricated nor was any data embellished.
2. Data was not misrepresented, especially not to peer review.
3. Data that didn't agree with the hypothesis was not left out to peer review nor to the public.

The third one may be a little confusing. The fact is that in this case, the other studies that were left out of the public eye initially merely did not provide definitive proof of a vaccine, but they *did* show that it was beneficial though. The best results, only, were part of the public announcement. It is very difficult to label this as scientific misconduct of the same vane we have been talking about. For sure, the far more responsible thing to do would have been to report all the results at once, even to the lay public. Given the enormous human toll that the HIV/AIDS epidemic has inflicted, *any* progress would be welcome good news. It is therefore difficult to know what the incentives would be for neglecting full disclosure. Whatever the case, there does not appear to be any negative effects from this lapse of judgment. However, it may erode the trust of the public who may rightly respond, "Hey, I thought it worked better than that!"

ANIMAL RIGHTS GROUPS

The previously mentioned zeal with which animal rights activists fight for their cause may also ultimately incite violence against researchers, both who conduct their research responsibly and those who do not. Thus, in cases where such violence was incited by unethical treatment of animals by researchers, it becomes a matter of personal safety, a very real issue of life and death in extreme cases.

BERNARD KETTLEWELL

In some cases, even after fraud (or terribly executed science) is confirmed, the scientific establishment chooses to hold onto the validity of the experiments in question. Such was the case with Bernard Kettlewell in the 1880s.[2] Kettlewell had a theory that lighter colored moths were easier to see by birds and thus were eaten more often, leading to a decline in the population of the lighter colored moths. Kettlewell performed an experiment and claimed that indeed the lighter colored moths were eaten more often, arguing that they were easier to spot, inferring natural selection at work. There were, however, several issues with Kettlewell's experiment. First, he nailed dead moths to the trees for the birds to feed on. Because the trees were dark, the lighter colored moths inevitably were more visible to a predator. Second, the moths in question were known to rarely land and rest on tree trunks, meaning that the location they were placed in during the experiment did not closely approximate real conditions. Finally, birds don't normally eat moths that are on the side of a tree. Despite these flaws in the design of the experiment, the experiments were still viewed as being valid by the scientific community. One must question why. One possible answer to this is even scientists can have their conclusions biased by what they want or expect to see. If these moths were a critical part of an ecosystem, then an experiment that argues for or against their survival or impending demise may have influenced the steps that society would have taken to protect these moths. Such influence could have far-reaching effects on society if misdirected. Let's not forget that this study was done over 130 years ago. This may account for some of our questions today regarding the quality of the study.

---

[2] www.neatorama.com/2006/09/19/10scientific-frauds-that-rocked-the-world/, last accessed 9/22/11.

THE ELECTROMAGNETIC FIELD AND HIGH-TENSION POWER LINES

Robert Liburdy, a cell biologist at the Lawrence Berkeley National Laboratory, was a leading researcher investigating the potential dangers of the electromagnetic field (EMF).[3] Until Liburdy began his investigation, there was no evidence that demonstrated an increased health risk due to EMF. This alone does not make it impossible for there to indeed be an increased health risk, of course. However, whenever something is reported that challenges previously held notions, the wise will look upon it with a certain degree of skepticism. Liburdy's papers claimed that the fields influenced the function of cells by disrupting calcium. It was later discovered that he had left out, manipulated, or misrepresented data such that they agreed with his pre-experimental notion that there would be such effects. This is clearly a violation of scientific conduct and is in fact quite different from other cases where the interpretation of the data is what was skewed by the pre-experimental notions. The deliberate alteration of data is a clear instance of scientific misconduct, of poor ethics, while allowing your judgment to be influenced by your prejudices is poor science.

FRACKING AND POLLUTION

Although not regarded as an instance of scientific misconduct, presently some of the controversy surrounding hydrofracking is emblematic of the influence that science can have over the public.[4] The public counts on scientists to present their data regarding the safety of such processes honestly and fairly. When we fail to do so, we mislead the public into directions that it otherwise may not have followed, or perhaps may not have prevented. One example of this with respect to fracking is Conrad Volz's findings that discharge from treatment plants that accept Marcellus shale waste water are a danger to public health. One of the treatment facilities responded by claiming that Volz was incorrect to compare the discharge water to drinking water standards. Volz however has been unwavering in his claims, even in the face of threats of legal action from the company. If either Volz's claims are accurate or if the company's claims are false, extreme damage could potentially be done to the communities near where the fracking is occurring. If Volz's (and other's) claims are exaggerated or even downright false, and fracking is discontinued, jobs, money invested, and a potential relief for energy would be lost if they caused the fracking to stop. If, on the other hand, the company's claims are false and the concerns that Volz points to in his study are true, then the process should be stopped immediately before irreparable damage can be done to the humans living nearby using the drinking water or any of the other delicate ecosystems that may be affected. The debate around fracking continues to this day and is at least slightly like drug testing in that some feel the economic and national security benefits of energy independence putatively gained by fracking outweigh the environmental risks that many fracking supporters claim are exaggerated, anyway. As with climate change, however, economic reasons used to tip the scales toward being beneficial are difficult to defend.

---

[3] www.sciencemag.org/news/1999/06/emf-researcher-faked-data, last accessed 6/23/18.

[4] https://public-accountability.org/2015/06/freedom-fracked/, last accessed 6/23/18.

OTHER CONSIDERATIONS

From time to time, we will see an advertisement for a product or service that we sense is just too good to be true. When it turns out to indeed be too good to be true, the perpetrator(s) is/are guilty of false advertising, a *crime* that carries penalties. Although it may be lower stakes than an automobile brakes' ability to stop a car, or a medical device's ability to perform, all fabrications of data in a research setting are, *in essence*, false advertising. Although there is currently no law that will inflict legal penalties on researchers who fabricate data in a research lab, in cases where the work is funded federally or by a state, the argument can easily be made that it should be. For sure, employees of federal or state labs who fabricate data are subject to criminal trials and you are unlikely to stay employed if you do it, especially repeatedly. Their wanton disregard for ethical behavior can not only potentially incarcerate an innocent person in the case of a crime lab worker but also ultimately call into question all his or her evidence and results, even cases where he or she was honest. This may have the result of releasing guilty persons. For example, Annie Dookhan, formerly of the Massachusetts crime lab, served approximately 3 years of a 3–5-year prison term for fabricating evidence that led to charges being dropped in more than 21,000 low-level criminal drug trials.[5]

Science also has an obligation to help the public understand itself and its goals. In 2016, Merck[6] released an ad that appears to pressure parents into making sure their children get the human papillomavirus (HPV) vaccine. According to Merck, the intention of the ad is to make parents better aware of the link between HPV and cancer later in the person's (the child's) life, along with the CDC recommendation to vaccinate at ages 11 or 12. By recognizing the important role that parents play in the health care of their children, it targets them for compliance.

One of the commercials pulls at viewers' heartstrings with a woman who has cervical cancer (as a younger, unvaccinated version of herself) asking her parents if they knew the risks of HPV and cervical cancer.

The response to this ad is divided, with some applauding its tactics at trying to compel (or to some, guilt) parents (in)to vaccinating their children. Those opposed often cite the popular calls against direct-to-consumer advertising.

Merck claims one of its goals is education since many (only ~50% according to Merck) parents are familiar with the link to cancer. Similarities between these ads and anti-tobacco ads, which were also graphic, have been made by Jeffrey Kahn, director of the Berman Institutes of Bioethics at Johns Hopkins University. The safety concerns of the Merck vaccine Gardasil cited by many only add fuel to the fire.

---

[5] "Annie Dookhan, Key Figure in State Crime Lab Scandal, Released From Prison", John R. Ellement, *Boston Globe*, **4/12/2016** and "Drug Lab Scandal Results in More than 20,000 Convictions Dropped", Jon Schuppe, *NBC News*, **4/18/2017**.

[6] Fierce Pharma. Merck Holds Parents accountable in new Gardasil ad compaign. Beth Snyder Bulik, 6/28/16. Washington Post, to your health, Do the new Merck HPV ads guilt-trip parents or tell them hard truths? Both. Laurie McGinnety, 8/11/16; Business insider. A shocking new ad is shaming parents for not giving their children this unpopular vaccine. Lydia Ramsey, 7/15/16; www.youtube.com/watch?v=sLB0MaY7luE, last accessed 6/23/18.

Some also wonder if the taboo nature of discussing a sexually transmitted disease (STD) with an 11- or 12-year-old is a contributing factor to parents' apparent unwillingness to address this disease and the vaccine that protects from it. Such would also cause some to have a negative response to these ads. One can't help but wonder if the response to this commercial would be different if it were the CDC, the FDA, or some other nonprofit organization who produced it.

Furthermore, it is worth noting that although the ad is quite dramatic and "in your face," the commercial in question (which can still be viewed on YouTube as of June 2018) does not mention Gardasil *at all*. It appears to be genuinely aimed at educating the public rather than advertising Merck's product, at least to the unbiased view of this author.

In fact, the tone of the commercial in question is very different from a commercial for Merck's vaccine, which highlights the patients (predominantly women) exercising their power to choose to get the vaccine.

Sometimes, public relations persons skew reality into a gray area of dishonesty. Edward Bernays,[7] an Austrian-American pioneer in the field of public relations and propaganda, was an absolute mastermind of controlling the public image for his clients. He has been called the father of spin and the father of public relations. He was also named by *Life* as one of the 100 most influential Americans of the 20th century.[8] Among his many "contributions" is the promotion of female smoking in the 1920s by branding cigarettes as "torches of freedom." Bernays also had a public relations hand in the Central Intelligence Agency (CIA) operations in Guatemala in the 1950s as part of defending the interests of the United Fruit Company's banana source. He also heavily promoted the health benefits of bananas, even placing them in the hands of celebrities to promote them. He very effectively used psychology and social sciences, going so far as to author a 1947 essay titled "The Engineering of Consent," a title that should make any free-thinking person uncomfortable. Even in 1992, the impact of Bernays's tactics was arguably still widely employed. In 1992, the International Food Information Council hired R. G. Clotaire Rapaille, a psychologist who advised them on ways to win public support for genetically modified foods. He made suggestions for words to use (e.g., bounty, children, diversity, future, generations, improved, wholesome, and others) and words to lose (biotechnology, chemical, DNA, manipulate, scientists, and others).[9] With these sorts of lists in mind, it is likely that those paying careful attention to propaganda of all kinds may be more able to see the myriad of ways we are manipulated daily or have our consent "engineered."

THE USE OF CELEBRITIES TO ENGINEER CONSENT OR SELL PRODUCTS

Everybody with access to either TV or the Internet has, without doubt, seen one of many commercials where some manner of celebrity is a spokesperson for some kind

---

[7] https://en.wikipedia.org/wiki/Edward_Bernays, last accessed 6/23/18.
[8] www.deseretnews.com/article/119956/LIFE-LISTS-20TH-CENTURYS-MOST-INFLUENTIAL-AMERICANS.html.
[9] www.sourcewatch.org/index.php/International_Food_Information_Council, last accessed 6/23/18.

of product or service. Some of these products and services involve a medical treatment of some kind. For sure, these tactics must have measurable effectiveness. The companies would not pay for and run these ads if they were not. But their purpose may be different or at least there may be multiple purposes, in cases where celebrities or otherwise famous persons are used as spokespersons. At the very least, for products such as a shoe that a famous athlete uses, a following and increased use can be garnered from these commercials since people (particularly young aspiring athletes) want to "feel like (*insert your favorite athlete here*)". The likelihood, however, that a medicine for high cholesterol, blood pressure, or erectile dysfunction will see a similar boost for this type of reason seems small however, given that many medicines have specific side effects or other interactions that govern who can use what. You also need a prescription to get such medicines; a doctor must give you "permission" to use this product, while anyone with money or a credit card can go get the shoes. In these cases (celebrities and medical treatments), one very important function the ads *may* have is to make some afflictions less taboo for a potential patient to bring up with his or her health-care provider. Achieving this sort of outcome will increase the number of people receiving potentially important treatments. As with other direct to consumer ads, anything that gets people talking to their doctors is a good thing.

Of course, in some cases, what is called disease mongering may be at work. Effectively, disease mongering is taking a normal life change and turning it into something treatable. These afflictions and the **true suffering** they cause are all not life threatening in the strictest sense, but reducing the quality of life of affected persons. "Ailments" such as baldness, wrinkles, erectile dysfunction, and low energy with increasing age (particularly in men) are all addressable (I won't use the word "cure" here) through some manner of medical intervention.

Broader discussion of this topic, including if such ads are unethical, if pharma should "waste" time and money on such treatments, if getting old is something we should go against nature and try to "treat," and many other topics related to this (not the least of which is direct-to-consumer advertising), is beyond the scope of this book. The interested reader is encouraged to look online for more.

SCIENCE AND THE ENVIRONMENT

It is difficult to say that science has a responsibility to the environment and not evoke images of science only working for environmental causes. To say that science has a responsibility to the environment is certainly **not** to say that the only worthy research is research such as renewable energy or pollution cleaning endeavors. Although if science **can** do any of these, it should because its responsibility to the environment goes further. Instead, to say that science has a responsibility to the environment is to say that, among other things, if there is a more environmentally friendly way to achieve the same scientific ends, we should follow that way, rather than one that is more damaging to the environment. It also means that science must take measures to limit or altogether prevent the release of waste products, particularly those which are hazardous to the environment or human health. When releases occur, the scientific enterprise is obligated to shoulder the costs and, when appropriate, the task of cleaning up the release.

Science also must be respectful of the limited resources the planet has. For example, many chemical raw materials (for plastics, other materials, and pharmaceuticals) come from oil, the same place that a significant amount of the energy on the planet comes from. Leaving out for now whether we should rely on fossil fuels *at all*, especially for energy, as the oil supplies dwindle, the chemical industry will eventually have a major crisis on its hands as the source for its raw materials and solvents dries up. Other forms of research are no different in their consumption of resources. We, as an enterprise, must do so responsibly but not in a way that takes the resources away from others. We should (and) do our part to find alternative sources of raw materials and energy.

Furthermore, the honesty of scientists must remain beyond reproach. The impact of dishonesty is felt by the public strongly in two areas: (1) pharmaceutical research and (2) environmental research. Although there are certainly others, these two areas are perhaps the most impactful when dishonesty is uncovered. Dishonesty, no matter where it is, erodes trust. If the trust in science is eroded too far, say because yet another pharmaceutical company is accused of covering up side effects, people will be less willing to take life-saving medicines. Although it was slightly different than the aforementioned scenario, the now retracted and discredited study by Andrew Wakefield that "showed" a link between the MMR vaccine and autism continues to wreak havoc on public health. The ramifications for scientific misconduct are real, but even more so when the research is of clear public interest as it is one of the few things we must all share—the environment. Instances of climate scientists fudging data sully public trust in the research enterprise and can cause laws regarding important environmental regulations aimed at preserving the planet to weaken or be removed. Even in 2018 there are people who are seriously claiming that the earth is flat and that there is a grand conspiracy where NASA and others are doctoring photographs to make everything look spherical. Although these persons generally lay on the fringes of society, that they have so little trust in the scientific enterprise is troubling. This is more than just gross scientific illiteracy. Such distrust probably cannot be resolved by education alone. In short, it is the fault of science itself for not better policing the scientists and avoiding misconduct, even if those misbehaving are a minority.

All of this says nothing of the instances when scientists interpret their results incorrectly. A nonscientific public can only tolerate so much of this before they lose faith in the enterprise and those trying to carry it out thinking it is clueless, untrustworthy, or both. A person with a lower scientific literacy will inevitably struggle to identify the difference between bad ethics, bad science, honest mistakes, and science working iteratively (and thus, well). It sometimes is difficult to convince a nonscientist, particularly with limited scientific literacy, that a mistake in an interpretation of results is not in and of itself unethical. Also, sometimes an incomplete study may be published, with the intention of publishing a follow-up afterward only to have the conclusions proven wrong by the same authors. There are perfectly good reasons for breaking up work like this, however, and a scientifically literate person will be proficient at identifying such cases. Other times, a researcher may be careless and have made inadvertent changes to a protocol he or she was following that he or she doesn't notice he or she made. This can in principle lead to publishing a protocol that doesn't reflect what the researcher actually did, though such an extreme example gets close to the gray area between bad ethics and bad science. All of these are easier to identify and understand with scientific literacy.

These, unlike the damage caused by unethical behavior, can be resolved by better education. Science is iterative in how it builds knowledge. Subsequent results may put previous results in a different light and cause a change in the conclusion. Also, as time progresses, so does technology. As technology progresses, the instruments we use in our laboratories to acquire data get better. The better our data, the better (i.e., more confident) our conclusions can be. Scientists and educators must do a better job at explaining this to the public, particularly while they are still in school. As the public understanding of how science works improves, their ability to discern the difference between bad ethics, bad science, honest mistakes, and science working iteratively will also inevitably improve and so will their trust in the scientific enterprise.

WRAPPING UP

All scientists have an obligation to behave responsibly toward society. This not only means that they must take care to not damage the environment and not mistreat animal (or human) subjects, but also that they must not report a study to the lay-public before the work has been verified, that is, vetted by their peers via peer review. Inevitably, lay persons will act upon the report, and the consequences of that could be disastrous, both from a health and safety aspect and a financial one. We therefore can only conclude that before a report of a scientific study is sent to *The New York Times*, it simply **must** be vetted by the scientific community. There is also a not so subtle responsibility to science itself in that changing publicized reports of scientific breakthroughs is a surefire way to lose public trust for science.

7 The Role of Government in Science and Scientific Misconduct

Equal to the power of the purse is that government agencies are the primary architect of the regulations governing the use of, especially, human subjects in research, although some of these regulations are internationally recognized. The primary role that the federal government (at least in the United States) plays in scientific misconduct is to control the purse strings, that is, the funding. The government has conducted investigations into allegations of misconduct, especially when it involves industries such as the pharmaceutical industry.

The federal government, or at least some of the funding agencies, also make rules regarding review boards and, to some extent, scientific misconduct. These days, any institution receiving federal funds must have an institutional review board in some manner to oversee particularly the research on human subjects. Now, the Offices of Sponsored Research are also nearly ubiquitous on college campuses, and they play an essential role in gaining grants and ensuring that universities meet the regulations set out by the federal and state governments. The Office of Scientific Integrity does, however, conduct some investigations into allegations of scientific misconduct at academic institutions.

Embargoes enforced against other nations, although this may seem unbelievable, have caused some scientific controversies. For example, the U.S. Treasury Department's Office of Foreign Affairs Control (OFAC) has caused the delay or prevention of some publications over the years. Furthermore, the American Chemical Society (ACS) and other American publishers have been prohibited from providing comments via the peer-review process and any editorial services to authors from Iran, Iraq, Libya, Sudan, Cuba, and North Korea in the past. The OFAC ruling further stated that manuscripts accepted by a journal may only be reproduced in exactly the form that they are received. Let's be clear about what it says; that is, if a journal is going to accept a publication from an Iranian scientist, it must accept the manuscript with any and all typographical and layout errors (which is different from the final, print layout) with which it is received in. This ruling included peer-review activities that would provide scientific feedback as well. This was eventually ignored by at least the ACS,[1] while the working to get the ruling overturned. In late 2004, the regulations were eased after a lawsuit. *In essence,* the federal government has tried to dictate to an independent publisher how and what to publish. This can absolutely be argued to be a usurpation of constitutional rights. However, a counterargument that

---

[1] www.ncbi.nlm.nih.gov/pmc/articles/PMC381079/, last checked 6/24/18.

constitutional rights are extended to individuals and not to organizations is certainly one that is logical. That being said, an environment where federal law inhibits the progress of science by forbidding productive communication between researchers is somewhat akin to the Dark Ages where the Church stifled science.

As if this wasn't enough, in late 2006, the ACS temporarily expelled 36 of its Iranian members along with one Sudanese member from its ranks amid concerns of violating OFAC embargoes administered by the Treasury Department. About a month later, the ACS reinstated these members with several restrictions. It was the ACS's expressed intent to lift these restrictions upon receipt of a license from OFAC, and this has subsequently taken place. These expulsions were not a minor public relations problem for the Society. It ignited a month-long "blogfest" and was mentioned (at its merciful conclusion) in June 2007 issue of *Physics Today*.[2]

However, are the government regulations that caused these issues examples of scientific misconduct, or does it *force* misconduct? Perhaps it is neither, but it for certain strikes at one of and perhaps the most critical foundation of modern scientific enterprise-free and honest discussion and collaboration. Regulations such as these hurt science not only by inhibiting or preventing discussion but also by preventing the scientist in our own country from benefiting or using the truly impressive work that may be done in these countries. It furthermore inhibits our scientists from profiting from their research in these countries. National security, for sure, is important, but perhaps there are better ways to ensure it than to get in the way of science and other scholarly work.

In August 2007, the U.S. Patent and Trademark Office passed new regulations that were drawn up by the Bush administration. These rules were immediately opposed by several companies, including GlaskoSmithKline (GSK). Because of these collective objections, the rules were put on hold, while the two sides fought in court. In short, the rules would have limited the number of claims that could be made in an individual patent filing. The goal of these rules was to help the office to reduce the backlog of unexamined patents and reduce the length of time it ordinarily takes to complete the application review. It is unclear that this change would have reached these ends because, in all likelihood, the number of applications would have increased under these rules, perhaps only making the matter worse. GSK argued, and this argument is indeed very sound, that this would have the effect of stifling innovation due to the fact that the applications are often constantly evolving on a drug candidate's pathway toward becoming a drug. Also, although this point was not reported to be a part of GSK's argument, these rules would have potentially caused an application to be less thorough and consequently, less likely to completely protect the interests of the filer(s). If true, this would logically result in a greater number of patents being circumvented, reducing the potential financial gains and likely ingenuity as well since the incentive can be taken as reduced, and thus less people will be willing to do it. Although the rules were rescinded in 2009, the entire controversy represents an unfortunate foray into science by the government and how, even when the government has truly positive intentions, its involvement can cause significant

---

[2] https://physicstoday.scitation.org/doi/10.1063/1.2754598, last accessed 6/24/18.

harm. When regulations governing science are authored by legislators without extensive input from scientists, this becomes a certainty.

According to the U.S. Copyright Office,[3] a copyright on a published work lasts the lifetime of the author plus an additional 70 years. Fair use is a concept often invoked in academic settings and permits the unlicensed use of copyright-protected work. The U.S. Copyright Office cites the following factors to be considered when declaring a fair use[4]:

1. Purpose and character of the use, including whether it is used for a commercial nature or for nonprofit educational purposes
2. Nature of the copyrighted work
3. Amount and substantiality of the portion used in relation to the copyrighted work as a whole
4. Effect of the use upon the potential market for or value of the copyrighted work

Other issues can also be taken into consideration on a case-by-case basis. In point 1, the courts are generally lenient in allowing educational uses of materials, although other factors may weigh heavily against a user. In point 2, a factual work being used is likely to be considered fair use, while creative work and unpublished work are not. In point 3, generally speaking, the lower the percentage of the overall work being used, the more likely it is to be considered fair use. Naturally, there are exceptions that courts have identified in individual cases. Finally, in point 4, if the use of the work is found to be unaffecting the market value/share (i.e., monetary value) of the work, it is likely to be found to be fair use.

Trademark protections are significantly shorter lived. According to U.S. laws,[5] trademarks initially last only 10 years. They can, however, be extended indefinitely as long as it is reissued. The United States Patent and Trademark Office further stipulates that in between the fifth and the sixth year of a trademark's lifetime, the registrant must submit an affidavit stating that the trademark is still in use. Failure to do so will result in the registration being canceled.

At least in the United States, the federal government also has the power to render decisions about patents. The government also makes decisions about trademarks and copyrights. Effectively, all things that are intellectual property (IP) fall under the government's purview. The Supreme Court of the United States has weighed[6] in on the patentability of genetic material. In the 2013 case, *Association for Molecular Pathology v. Myriad Genetics, Inc.*, the Supreme Court held that as products of

---

[3] www.copyright.gov/title17/title17.pdf, last accessed 6/24/18.
[4] www.copyright.gov/fair-use/more-info.html, last accessed 6/24/18.
[5] www.registeringatrademark.com/length-trademark.shtml, last accessed 6/24/18.
[6] Supreme Court October term, 2012, syllabus number 12-398. *Association for Molecular Pathology v. Myriad Genetics, Inc.* Business wire "Ambry Genetics defends appeal by Myriad Genetics to restrict Marker for Breast-ovarian Cancer genetic testing, 12/17/17. Washington Post, Timothy Lee, You Can't patent Human genes so why are genetic testing companies getting sued? July 12, 2013. The New York Times, Andrew Pollack, Myriad Genetics ending dispute on Breast Cancer risk testing. January 27, 2015.

nature, human genes cannot be patented. However, if tests or other methods that use or in some way analyze human genes can be patentable. Myriad continues to maintain such use patents in a lawsuit against Ambry for the latter's use of certain technology Myriad claims is theirs. Although in the 2012 case *Mayo v. Prometheus*, the Supreme Court ruled similar patents that were invalid, in late 2014, Ambry genetics defeated the lawsuit. The Court of Appeals for the Federal Circuit upheld the decision of a Utah Federal District judge that ruled the claims in question ineligible for patents. The courts held that since Myriad did not create or alter the genes, they were not eligible to be patented. The court said Myriad developed methods to find that the genes using well-established industry techniques were not sufficient to grant them a virtual monopoly on testing that identifies such genes. By early 2015, Myriad had opted to drop the matter and begin to pursue settlements with its competitors.

Genetically modified organisms, on the other hand, are patentable under current (as of this writing) U.S. law.[7] So what is the difference? There are several qualities something must have in order to be patentable. The process, machine, manufacture, or composition of matter must be new and useful. As an outlandish example, this means you cannot patent a T-Rex trap. It would fail to meet the useful requirements. Improvements and all other aspects of an invention must be novel and nonobvious to someone learned in the art. With the publication of the human genome project, anything regarding human genes as they exist in nature is considered nonnovel. Genetic therapy, which creates changes to genes as they exist in nature, is patentable since they constitute a change to a composition of matter. Genetically modified organisms fall under a similar category since they also contain a change in the composition of matter that would not exist otherwise. Arguing that evolution could have brought about the same change currently does not hold enough water to invalidate a patent. I am not aware of any instances where a genetically modified gene was subsequently found in nature.

It is entirely possible, perhaps likely, that as these industries and treatments mature, the laws that govern such patents will change in favor of making more products patentable. There are no indications, however, that this will happen any time soon.

The issue of funding controversial research such as stem cell research is also one that, unfortunately, the government cannot help but be stuck in though in this case, which is certainly not the government's fault. Those who oppose it usually do so from a religious perspective, to the great anger of those who support it, but nonreligious arguments can certainly be made against it as well. The supporters claim that such grounds (religious) for opposition are a violation of Church and State separation guaranteed by the First Amendment of the Constitution. As for the initial regulations restricting funding for embryonic stem cell research put in place by President George W. Bush, in 2009, President Barak Obama overturned the regulations set forth by the Bush administration.[8]

Although not related to our federal government, and not necessarily related to misconduct, other governments have also caused quite a stir. For example, in the

---

[7] https://ghr.nlm.nih.gov/primer/testing/genepatents and www.genome.gov/19016590/intellectual-property/, last accessed 6/24/18.

[8] www.nature.com/news/2009/090309/full/458130a.html, last accessed 6/24/18.

past, German authorities have not allowed scholars educated in the United States to use the title "Dr." without special permission.[9] Today, we look at such rules as being arcane and downright bizarre. And, indeed, this controversy has been quashed. One of the cases that attracted a good deal of attention was that of Ian Baldwin. Dr. Baldwin received summons from the city of Jena's criminal investigation department claiming he was a suspect in an investigation into the abuse of a title. He faced a fine and up to 1-year imprisonment for violation of the law, with its origin in 1939, which states that only holders of a doctorate from European nations were permitted to use the title, "Dr." American-educated PhDs were not permitted to use this title. The directors of the Max Planck Institute (Baldwin's employer) retained legal counsel and fought the law. Since the eruption of this controversy, State education ministers have met in Berlin and have decided that American PhDs are now (finally) legally permitted to use the title Dr. in Germany.

WRAPPING UP

Although its intentions are hopefully always good, it often does nothing but make matters worse when the government gets involved in science (the NIH, NSF, and NASA not withstanding though, these federal entities **must be** run by scientists, not lawmakers). The clearest case of this is the OFAC rulings previously discussed. Their intent was to prevent countries belligerent to the United States from benefiting from services provided by the United States. The effect that they instead produced was one that challenged one of the very essences of science—uncensored and unrestricted communication between scientists. The one case where federal intervention is appropriate is in matters of federally funded research tainted by scientific misconduct. Even this, however, comes with a stipulation: scientists and **not** lawmakers should make up the "jury" in the rulings of fraud. From there, the federal government should be allowed to impose any legal penalties it sees fit to the offending scientists *in addition to* any of the penalties the scientific society imposes.

---

[9] www.spiegel.de/international/germany/0,1518,540459,00.html, last accessed 6/24/18.

8 Is There Research That Shouldn't Be Done Because It Is too Dangerous?

With the increased concern of global terrorism, especially bioterrorism, there is increasing scrutiny regarding research in pathogenic entities.[1] Much of the present-day concern started with the anthrax cases in the United States in late 2001. Specifically, this highlighted a potential vulnerability whereby terrorists (domestic or foreign) could weaponize biomedical research. In principle, such individuals could use this research as a cookbook of sorts to fabricate a bioweapon. As a result, research and the subsequent reporting of it are under increasing scrutiny from funding agencies and all manners of worldwide defense organizations. Although some of the restrictions have been self-imposed by the research community, others have come from the funding agencies. There have also been some calls for relegating such research under the umbrella of the Department of Defense or Homeland Security (at least in the United States), who are able to fund confidential/secret research, unlike the National Institutes of Health (NIH), which uses an open access model to its funded research. The focus of this to date has centered around research on the flu virus, although there are certainly other examples too. This is because the flu has the potential to reach pandemic levels, which can lead to a worldwide catastrophe. This is more than just hyperbole. The Spanish flu of 1918–1919 killed tens of millions of people.[2] Although modern medicine would most likely result in less deaths today, such a reduction may very well be offset by the increase in global travel, compared to a century ago. This would aid in its spread and set the stage for a truly epic medical catastrophe that even modern medicine may not be able to keep up with. It is worth noting, however, that to truly reap the most protection from such research regulations, there must be a coordinated effort across the globe. Furthermore, some publishers have taken it upon themselves to not publish, for example, a study that tinkers with anthrax to make it more deadly. Thus, a two-pronged approach—one that seeks to regulate what kinds of research are done and the other that modulates what is reported—is undertaken.

The Department of Health and Human Services (HHS) has established an updated framework (2017)[3] to guide funding decisions for research involving enhanced

---

[1] https://en.wikipedia.org/wiki/2001_anthrax_attacks.

[2] https://Virus.standford.edu/uda.

[3] U.S. Department of Health and Human Services: Framework for guiding funding decisions about proposed research involving enhanced potential pandemic pathogens, 2017.

potential pandemic pathogens. This framework defines a potential pandemic pathogen as the one that is both highly likely to be capable of wide and uncontrollable spread in humans and likely to be highly virulent and likely to cause significant morbidity and/or mortality in humans. Enhanced potential pandemic pathogens result from enhancement of transmutability and/or morbidity.

It is important to note that studies that investigate how the virus responds to drugs or characterizes naturally occurring viruses would **not** receive this extra review. To be eligible for funding, a multidisciplinary panel of federal experts with backgrounds in public health, medicine, security, science policy, global health, risk assessment, U.S. law, and ethics must find that the following eight criteria are met:

1. The research has been evaluated by an independent expert review process (whether internal or external) and has been determined to be scientifically sound.
2. The pathogen that is anticipated to be created, transferred, or used by the research must be reasonably judged to be a credible source of potential future human pandemic.
3. An assessment of the overall potential risks and benefits associated with the research determines that the potential risks compared to the potential benefits to society are justified.
4. There are no feasible, equally efficacious alternative methods to address the same question in a manner that poses less risk than does the proposed approach.
5. The investigator and the institution where the research would be carried out have the demonstrated capacity and commitment to conduct it safely and securely, and have the ability to respond rapidly, mitigate potential risks, and take corrective actions in response to laboratory accidents, lapses in protocol and procedures, and potential security breaches.
6. The research's results are anticipated to be responsibly communicated, in compliance with applicable laws, regulations, and policies, and any terms and conditions of funding, in order to realize their potential benefit.
7. The research will be supported through funding mechanisms that allow for appropriate management of risks and ongoing federal and institutional oversight of all aspects of the research throughout the course of the research.
8. The research is ethically justifiable. Non-maleficence, beneficence, justice, respect for persons, scientific freedom, and responsible stewardship are among the ethical values that should be considered by a multidisciplinary review process in making decisions about whether to fund research involving potential pandemic pathogens.

The framework goes on to outline the responsibilities of the HHS and the funding agency for this sort of research.

Other research is also controversial; perhaps the most well-known is genetic therapies or, more generally, genetic research. For one, there are (and likely will forever be) arguments over when life starts and someone is human that fuels the opposition to especially embryonic stem cell research. To some, life starts at conception, and so embryos ought to be owed all the protections of any other human being in clinical trials. The merits and demerits of this stance will not be considered here.

In 2016, the NIH put into place new regulations[4] that effectively lifted a moratorium on animal–human chimeric research. In this research, human stem cells are added to animal embryos. Goals of such research would be to study human development, generate disease models, or potentially grow human organs to be used in transplantation, particularly with pig–human or sheep–human chimeras. The ban was intended to continue for any chimeric animals that may carry human eggs or sperm and tighten to include adding stem cells at any stage in primate embryos. Chimera experiments that would be permitted (at least after review by a NIH steering committee of scientists, ethicists, and animal welfare experts) are adding human stem cells through the gastrulation stage and introduction of human stem cells into the brains of postgastrulation mammals, except rodents that will not require extra review. The steering committee shall consider factors such as the type of human cells, where they wind up in the animal, and how the animal's behavior or appearance may change as a result.

It is fair to ask why such research would be undertaken, both in the context of the chimeric research and the gain-of-function research for pathogens. The chimeric research has been briefly addressed already, but regarding the gain-of-function research, if it is possible, somebody, somewhere, may also do it but with sinister motives and intentions. It is therefore good preparedness and defense to be prepared to counter such actions, and the only way to develop countermeasures (e.g., treatments) is to research such pathogens. Another answer is that although nature, to our knowledge anyway, has not developed the mutation the research directs yet, it doesn't mean it will not ever. Once again, this research allows for better preparedness for this sort of pathogen to be encountered, whatever the source.

It is worth noting that in 2017, the U.S. FDA approved its first genetic therapy. The treatment, called Kymriah, is used to treat children and young adults who have a type of leukemia. More such treatments are no doubt on the way. Some other treatments that are currently under investigation are as follows[5]:

- Cancer diseases (64.6%)
- Monogenic diseases (10.5%)
- Infectious diseases (7.4%)
- Cardiovascular diseases (7.4%)
- Neurological diseases (1.8%)
- Ocular diseases (1.4%)
- Inflammatory diseases (0.6%)
- Others (2.3%)
- Gene marking (2%)
- Healthy volunteers (2.2%)

---

[4] www.sciencemag.org/news/2016/08/nih-moves-lift-moratorium-animal-human-chimera-research, last accessed 6/24/18.

[5] http://abedia.com/wiley/indications.php, last accessed 6/24/18.

All of these (except the healthy volunteers, which is necessary for safety testing) have one thing in common: they are all intended to be therapeutic; none are cosmetic or quality of life. Naturally, genetic modifications are not the only controversial topic being researched. Although weapons, especially weapons of mass destruction, will likely be something objectionable, we forever live with that fact that artificial intelligence (AI) in all its manifestations (e.g., cyborgs, the self-driving car, and the predictive power of autocomplete on search engines and autocorrect) is slowly becoming ubiquitous. Although there are currently no FDA-style government agencies regulating such research, there are an agreed upon set of principles regarding AI research called Asilomar AI principles. These principles were developed in part by AI/robotics researchers.[6] The list of 23 nonbinding principles was developed to try to ensure that AI would not be misused and lead to a situation fit for an apocalyptic Hollywood movie. The principles are categorized into three groups:

1. Research issues (principles 1–5)
2. Ethics and values (principles 6–18)
3. Longer term issues (principles 19–23)

Adherence, by all, to these principles should ensure that AI benefits humankind, provided, of course, nobody hacks the system to corrupt its proper function. Perhaps the inevitable hacking should be enough to dissuade such research. To date, it is not and therein lies a major issue, potentially. The principles are as follows:

1. Research goal
2. Research funding
3. Science-policy link
4. Research culture
5. Race avoidance
6. Safety
7. Failure transparency
8. Judicial transparency
9. Responsibility
10. Value alignment
11. Human values
12. Personal privacy
13. Liberty and privacy
14. Shared benefit
15. Shared prosperity
16. Human control
17. Non-subversion
18. AI arms race
19. Capability caution
20. Importance
21. Risks
22. Recursive self-improvement
23. Common good

Taking a few and expanding on them:

1. Research goal

The goal should be to create beneficial intelligence, rather than undirected intelligence. This means that the AI system should only benefit its creators and users. A system with undirected and continuously expanding intelligence will eventually become impossible to control and thus a real threat.

---

[6] https://futureoflife.org/ai-principles/, last accessed 6/24/18.

10. Value alignment

The goals and behaviors of an AI system should align with human values throughout its operation. While this absolutely begs the question of whose values, two of the other principles can combine to advance this. Where value alignments specifically mention human values, which are cited as a principle of its own, is defined as being compatible with the ideals of human dignity, rights, freedoms, and cultural diversity. In short, AI cannot make us its slave (oddly, we can make it our slave), nor can it be used as a vehicle for "ethnic cleansing."

16. Human control

This principle states that AI should only perform objectives or decision-making for a very limited set of instances that the user (human) designates. This is important because if left unchecked, AI systems will control and execute more tasks and decision-making as it self-improves. It is therefore essential that it only be given the capabilities and permissions to perform a limited set of tasks. Otherwise, we can lose control of the very system we create as it simply does whatever it determines is the best action.

18. AI arms race

This principle in short states that lethal autonomous weapons should be avoided. Personally, the use of the word "should," rather than *shall* or *must*, is a little worrisome, but perhaps I am reading it too literally. In any event, that it would be a catastrophe if AI gained the ability to furnish and utilize arms, especially against humans, is a grand understatement.

22. Recursive self-improvement

This principle states that the degree to which AI can self-improve and self-replicate must be strictly controlled. For sure, we want an AI system to improve itself as the better it is, the better the benefits we reap will be. Autocorrect, voice recognition, and search autocompletes all get better with time because they learn our patterns. This makes the product better, but it is only possible because it is programmed to do so. The ability to self-replicate is also important but even more important to restrict. This improves humankind in that if the benefits are more widely dispersed, this would be better. Unchecked self-replication, however, could eventually (at least in concept) result in the AI system increasing in number so greatly that there are too many to control. The self-improving nature too must be monitored, although it may be harder to envision a doomsday scenario, since **no matter how** intelligent we are, a computer has (at least computationally) virtually limitless intelligence. We therefore are best served to force a limit, lest "they think they're in charge."

WRAPPING UP

Research can be dangerous. This is from multiple perspectives, not just from the carrying out of the research (i.e., the danger inherent in actually using the chemicals or processes) but also from the implications of the research. How the research product is used, also, could be a matter of grave danger for all humanity. It therefore leads to the question: "We *can* do this, but *should we?*" Although a necessary question, in my opinion, I also do not think it can be asked lightly. This is a very slippery slope and can easily become repression of science.

9 (How) Can We Prevent Scientific Misconduct?

Unfortunately, in the absence of some sort of worldwide change in human nature, ethical violations are probably not completely preventable. Terrible draconian penalties such as revocation of tenure, immediate termination, or even being sentenced to a multiyear publication or grant submission ban on the offending author may help but the potential for effectiveness is at best questionable. However, it is very likely that through proper education, the number of cases can be reduced—a major goal of this book. Such a reduction can be affected in at least two different ways: (1) Previously, what is now understood to be an ethical violation was not known prior and will not take place now. (2) There will now exist a heightened awareness such that a potential accidental perpetrator will catch himself or herself before he or she commits an ethical violation. Additionally, all institutions have some form of office of research integrity, whose job is to verify that its scientists are not committing scientific misconduct. Usually, investigations would begin in such offices if charges of fraud are made.

One particular draconian penalty may be to mandate that researchers who commit fraud of any kind would have to pay back a percentage of any grants that funded the affected work. This would be fair as it represents a misuse of public funds. A flaw in this is if it was a "low-level worker," does the principal investigator (the PI, the boss) pay or the worker? Presumably, the percentage would be determined by the severity of the foul and the number of years left on the grant or perhaps all of the grant amount, if the grant is over and funded ultimately fabricated work. Of course, it is highly debatable whether or not such draconian penalties would be effective deterrents. A prime example of the potential for extremely draconian penalties is the death penalty, which has debatable effectiveness at preventing capital crime. Now, nobody would suggest that researchers who commit fraud should be put to death. However, if this (death) is not an effective method of deterring capital crime, there is no reason to suppose that paying some percentage (or all) of a grant back would be effective at preventing fraud. A brief exploration into preventing each type of misconduct may be informative nonetheless.

INTENTIONAL NEGLIGENCE IN ACKNOWLEDGMENT OF PREVIOUS WORK DONE

The only really effective way to prevent or reduce this violation of proper scientific conduct is via the peer-review process, where it potentially can be caught and resolved before publication. With diligent peer reviewers, work that is neglected can be included in the manuscript. Of course, the editor's cooperation is essential for this to succeed. If a peer reviewer thinks that something should be referenced, the editor should take this into consideration. Safeguards must be put in place, however, that prevent reviewers from abusing this for their own benefit. Even with anonymous peer reviews, the editor always knows who the reviewers are. For certain, since reviewers are ordinarily experts in a field, they may occasionally review a paper that ought to acknowledge some of their own work in the area. However, if a reviewer makes a chronic habit of making such recommendations, the editor should put an end to this and either not use the review or ask for an additional review from a perhaps less-biased reviewer.

DELIBERATE FABRICATION OF DATA

Wholesale prevention of data fabrication is likely impossible. Fortunately, as discussed earlier, it is almost always found out, at least eventually. Although, in truth, there is nothing to prevent a researcher from simply claiming that mistakes were made, especially if the fabricated data is written in a laboratory notebook, the first place an investigation would look. One tactic that *may* reduce the incidence of data fabrication would be to remove or reduce the incentives (i.e., awards) for great scientific achievements. This would be completely unreasonable as, under this type of system, the greatest minds will have no clear reason to enter into a scientific career. The only other tactic that *may* be effective would be to increase the penalties for verified instances of fabrication of data, exposing the offending researchers to penalties such as lawsuits charged by people directly harmed by the fabricated data. Once again, however, the effectiveness of such penalties is by no means ensured. One immediate complication that arises under this scheme however is just who is responsible? For example, if a student does a masterful job at deceiving his or her adviser, who should pay the price? Is the professor, at the end, chiefly responsible for the quality and integrity of the work? Or is it *always* the person who actually committed the fraud? (Though, frankly, suing a graduate student may be the worst way to get money ever conceived.) Whatever the penalty is, it must be "payable" by those charged. By this, I mean that few people, especially graduate students, would be capable of paying back a grant. Furthermore, since in most cases, no criminal activity has occurred during scientific misconduct, to have the penalty for it being even equally severe to a *bona fide* crime is likely to be considered cruel and unusual punishment.

DELIBERATE OMISSION OF KNOWN DATA THAT DOESN'T AGREE WITH HYPOTHESES

Omitting data that does not agree with the hypothesis is quite different and, in fact, quite easy to address. As discussed earlier, this transgression may border defensible. There *is* a way to prevent this ethical violation from occurring. Previously, we supposed, this usually occurs because aberrant data is often the cause for rejecting a paper via the peer-review process. This is a practice that must simply be ended in my opinion. Work is more complete when the limits are presented, and to reject work that discusses those limits is unscientific in every way. Part of the essence of science is knowing not only what works well but also what doesn't work. That way, the all-important WHY can be explored and perhaps the limitations can even be overcome. Efforts simply must be made to change the paradigm away from this type of review. If the (even if only perceived) consequences of incorporating such data were removed, researchers would probably be less likely to omit data. Perhaps, it would be most appropriate to begin this culture earlier in a student's education, even at the high school level, putting larger premiums on explaining the result he or she observes during a laboratory exercise. It may be worth deliberately installing experiments with mediocre (at best) success or even some "planned failure" to get students "comfortable with" incorporating such results. That being said, we cannot have a wholesale exodus from caring about positive results. A student with a 13% yield on a chemical reaction where the rest of the class gets >90% **must** receive some sort of grade penalty; similarly, research that leads to better protocols are better than those that are less good, no matter how logical and eloquently articulated the explanation is for why something may be inferior.

PASSING ANOTHER RESEARCHER'S DATA AS ONE'S OWN

Similar to the fabrication of data, this likely can not be truly prevented; however, its reduction warrants an expanded discussion. It may be possible to reduce this violation more than the fabrication violation through some education. For example, it is really not "competition" and "scooping" your rival researcher if you take his or her work, quickly reproduce it so you can say that *you* actually did it, and then try to publish it as your own. This is simply not right because the work was usually presented to you in good faith as a private communication in a department seminar, or in an article/grant you are reviewing. With more education, perhaps more researchers can learn that this is an unacceptable practice in every way—it is **not** good healthy competition. Again, to be fair, this exact type of instance is very rare. Likewise, it is against the norms of science to present previous work done by both your laboratory and others and not accurately and properly cite this work. By increasing awareness of self plagiarism, this brand of scientific misconduct can be reduced.

Other than increasing education and awareness of this violation of scientific code and trust, or extremely draconian penalties, the best way to prevent this may be to increase the number of reviewers, thereby increasing the likelihood that someone will notice. This however is fraught with impracticalities. As stated before, the peer reviewers are nearly all unpaid volunteers. Asking more reviewers to give up what is already precious little time is unreasonable. Furthermore, once again, it is not the intended job of peer review to ferret out these violations.

PUBLICATION OF RESULTS WITHOUT CONSENT OF ALL THE RESEARCHERS

Preventing this is truly quite easy. Every author should be sent a copy of the paper prior to submission and given a deadline by which to read it (at least 2 weeks). In these cases, there is enough time to read the paper and voice any objections/consent or have there be consent by way of not raising objections. Some journals require a statement from each author, while others contact all of the authors. All journals should adopt these sorts of practices. One may think that another possible route would be to have all publications be submitted to journals through the office of sponsored research, which often handles grant submissions at academic institutions. With this sort of administrative oversight, these issues can be much more easily addressed prior to the submission of a manuscript, when there is still time to resolve them easily. There are at least two issues, however, that make this an unworkable process. First, such offices are not set up to handle the workload that would result from this. Second, often, collaborations are researchers at different institutions. The question that would then inevitably come up is "whose office handles this?"

FAILURE TO ACKNOWLEDGE ALL THE RESEARCHERS WHO PERFORMED THE WORK

In cases where someone is innocently left off a paper, the only way to assure its prevention seems to be to encourage laboratory directors to better keep their records and simply be more diligent. None of us are perfect; however, so eliminating by this method is likely a long shot. However, social and professional networks online seem logical places to start. In cases where it is intentional, due to a personal conflict, there may be no prevention. Human nature causes us to seek revenge, and this is one method of professional revenge in science. In other cases, where authorship is contested or argued, it would be most helpful and logical to discuss the parameters for authorship at the outset of a project or at least prior to preparing a manuscript. This is, of course, a time-consuming endeavor, but it would be very effective at reducing controversies.

To resolve this, and the previous violation, at least with respect to patents, all patent applications are completed with the assistance of a lawyer. This provides an ideal mechanism by which all authors can be included rightfully. Indeed, all patent applications require the permanent address and signature of all of the authors/inventors. Of course, this only includes the authors the PI lists as coinventors on the patent. There is, however, something that can be done. The attorney(s) can (and perhaps should) interview *all* of the workers in the lab(s) involved in the patent to ensure that nobody who feels that they contributed has been left off the patent. If someone charges that he or she does feel this way, there is time to investigate and resolve such claims before the patent is ultimately filed. This, for sure, may cost a fair amount of money for the increased time demanded of the counsel. However, since litigation that accompanies controversy around patent authorship is always significantly more expensive than this, it may be worth the investment.

Similar regulations can be put into place regarding journal publications as well, although certainly not with lawyers. A university or other institutional offices such as the office of sponsored research, or some intradepartmental committee, could in principle at least require a statement from all of the members of the research group attesting to the author list on every publication. At the very least, the PI could create a paper trail that documents that nobody in the research group feels they have been left out. A diligent PI can even collect signed copies of notebook pages and only incorporate data from these pages. This would immediately allow him or her to attribute each experiment to an individual researcher and the list would frankly make itself. It may also help to assuage the concerns of objecting authors who feel as if they contributed more. Such a process would be workable for even the most prolifically productive research groups and institutions.

Although not formally a preventative measure, per se, something can be done if this issue is uncovered after a paper is in publication. The same can be said for author order disputes. A corrigendum (or correction) could be filed and published that fixes the wrong credit though, sometimes, it is retracted instead.

CONFLICT OF INTEREST ISSUES

For sure, the fast and draconian way to fix this is to mandate that no academic lab directors are allowed to have any financial interest in their work. Clearly, this is an inappropriate solution because if a lab or researcher develops something that makes, for example, a pharmaceutical company millions of dollars, the academicians deserve something for their role in the discovery beyond a pat on the back. This sort of mandate would also undoubtedly have negative repercussions on books. Most academic texts are written by professors, and placing oppressive restrictions on their production would greatly harm academic progress. Faculty persons should also be allowed to use their own personal money to start up a company based on their research to fund and subsequently profit from their research, provided the necessary institutional regulations are also followed. Without inappropriate restrictions, the best thing to do is to have the university or other administrating body hire an independent mediator; failing that, the researcher can ask a trusted colleague to point out when his or her personal judgment is superseding his or her professional judgment and most universities, in fact, do. Most institutions also have rigorous records that contain declarations of conflicts of interest for monitoring and self-auditing purposes. These are often enough to curb instances of conflicts of interest since they declare specifically that they may exist, resulting in everyone watching.

REPEATED PUBLICATION OF TOO SIMILAR RESULTS

This is one that the peer reviewers and editors must team up to address. With reviewers and editors across journals working together, duplicate manuscripts would be rejected and prevented. Doing so completely is unlikely and maybe even impossible due to the volume of work being done, but for sure, this can at the very least be reduced. One potential solution may be to select a reviewer whose sole task is to check if any similar work has previously been published. This can be done quite simply with a handful of searches on SciFinder Scholar® or some similar database. However, once again, this is not the intent of peer review and may not be an appropriate use of resources as a result. Plagiarism detection software is also increasing in its sophistication and efficiency. Modern cheaters can only be stopped with modern tools and the publishers are catching up.

BREACH OF CONFIDENTIALITY

Like other violations mentioned above, the only likely way to prevent or reduce this ethical violation is to make the penalties as harsh as possible. Once again, however, the potential for success is questionable. An alternative may be to reduce the incentive to commit this breach for the offender, but even small benefits will be appealing enough to some.

MISREPRESENTING OTHERS WORK

Like fabrication of data, preventing this is probably not possible. The best we could hope for is to catch them sooner, and since, like other examples, it involves increasing the demands on peer review, it is unlikely to be reasonable. There is a chance that if peer review were to stop rejecting papers for bad results, one of the incentives to do this would be reduced, but this is probably a long shot since there are multiple incentives for this violation occurring.

EDUCATION IS KEY

One way that scientific misconduct could be reduced is to better create a culture in science education that it is an unacceptable practice during the students' education. This can potentially be done by repeatedly exposing students to the topic even to the point of it being a module in every one of their science classes. Such an approach, especially if it starts in high school (at a more basic level of course), will demonstrate to young scientists that there are certain standards and expectations on their behavior that they must rise to if they want to be a part of the scientific community. These behavioral expectations for proper conduct and for healthy collaborations for that matter must be demonstrated to students repeatedly and consistently, and I would argue globally. Such a coordinated effort in science would not be easy among diverse persons who already struggle to fit "all the important topics" into their courses. Still the potential benefits are worth it. With the current distrust that the public seems to have for science, it is not hard to see a second Dark Ages coming. This one, however, would be ruled by pseudoscience, rather than the church. The loss of trust in science is *this* critical, and we, scientists, must not let it happen.

WRAPPING UP

Unfortunately, overall prevention of scientific misconduct is simply never going to happen. The best that we can reasonably expect to do is catch it more quickly. In all but the cases that involve fabricated or omitted data, this can likely be done even if doing so would be an enormous order. More likely, however, there will always be bad apples who will try to beat the system. Failing lifetime bans for even the smallest misconducts (which again may not be effective), it will probably continue in some form or another.

Part B

Case Studies

10 Case Studies

Unlike the examples covered previously, for which it was clear that misconduct occurred, and penalties were levied, some of the cases that follow in this section are much less clear-cut. It is the intent behind this presentation that these cases stimulate conversation (and perhaps even lively debate) about whether misconduct has taken place.

DARWIN AND WALLACE

Summary: Charles Darwin, credited with the discovery of evolution, was actually not alone in this discovery. His contemporary, Alfred Wallace, is forgotten to all except a learned circle. Their papers, presenting their findings, were jointly presented at the Linnean Society in Burlington House, Piccadilly. Wallace's contribution is so forgotten, which is evident in everyday colloquialisms such as Darwinism, social Darwinism, and the sarcastic Darwin Awards.

What happened? It would not be a stretch to say that Wallace did essentially what Darwin did, only in Brazil (and thus logically, with other specific animals). One of the major differences in their interpretations of the results is that while Darwin pursued survival of the fittest as the impetus for evolution, Wallace pursued environmental forces as its incentive. Looking at these two explanations with modern eyes reveals little difference between the two since the fittest survive in their own specific environment. But, during their time, these were different. Furthermore, Wallace performed his studies nearly two decades after Darwin performed his. To be fair, Wallace suffered a major setback when nearly everything was lost in a ship fire, contributing to his delay. So, what happened that their studies were released together? One must keep in mind that this was the mid- to early 1800s. The feelings about creationism then were *far* more intense than they are, even today. During that time (in Judeo-Christian societies anyway), believers in anything but the biblical view were literally considered heresy. With the Church (especially in Europe) wielding so much power, being branded a heretic then was one of the worst sentences one could incur. Therefore, Darwin's hesitation and his preference to be *absolutely sure* are quite understandable. Darwin received the confidence he needed from a letter sent to him by Wallace (whom also shared bird samples with Darwin) in which Wallace described his theory. This theory allegedly came to him (Wallace) in a fever-induced dream. This letter was also not the first time the two had communicated. He sent this letter to Darwin, requesting Darwin review the paper, and if he (Darwin) thought it is worthy, it was passed on to Charles Lyell. To Darwin's horror, this theory was very similar to his own. What happened next was, at the very least, curious. Darwin wrote to friends Joseph Hooker and Charles Lyell, lamenting that someone may get the credit for his discovery. It should be noted that Darwin was a far more respected and renowned researcher than Wallace. Had the theory only been presented by Wallace, it may have done more damage to the theory than good. Hooker and Lyell determined (and their friend Darwin's interests were almost undoubtedly an incentive for them) to present both theories together.

Wallace, for his part, was allegedly pleased with this outcome. It gave him significant credit (even if some of it is lost in the translation in modern times) and effectively inducted him into an "inner circle" of sorts containing the greatest and most respected scientific minds of the day. He later went on to more or less father the field of biogeography. Meanwhile, Darwin proceeded to author his epic *Origin of Species*. Perhaps these last two points combined are the *real* reason Darwin is the receptor of most of the popular credit. Darwin went on to continue to contribute to the field, while Wallace pursued another, albeit related one. Wallace also did not even once publicly cry foul against Darwin, Lyell, or Hooker. This has not stopped some from writing treatises about Darwin cheating Wallace out of the discovery, even incorporating illegitimately some of Wallace's work into his own. These claims have been found by nearly all historians of the field to be completely incorrect and unfounded.

Resolution: It is always possible that some unethical behavior may have occurred in cases such as this. The evidence argues strongly against these claims. Wallace was given due credit at the time. The primary reason Darwin gets nearly **all** the credit today is likely the fact that he continued to contribute to the field, while Wallace embarked upon other pursuits. Furthermore, Wallace was never once reported to have taken issue with the outcome of the correspondence with Darwin. Finally, if Darwin wanted to, he could have just sat on Wallace's letter and never passed it on to Hooker and Lyell. The fact that he did not do this may stand as the most compelling piece of evidence in his favor. Furthermore, since 1908, the Darwin–Wallace medal has been awarded by the Linnean Society of London. Initially, this award was issued every 50 years, starting in 1908. More recently (2010), the award is issued annually. This suggests that, at least since the award's inception, 50 years after the work of Wallace and Darwin was reported, the *scientific community*, whose opinion is perhaps the one that really counts, recognized the contributions of **both** men. That popular culture has all but forgotten Wallace is not Darwin's fault.

Questions to ponder:

1. Was Wallace wise to contact Darwin and divulge what he did?
2. Did Darwin behave unethically?
3. Did Hooker and Lyell behave unethically?
4. How can we better recognize Wallace's contributions moving forward? Should we?

SOURCES

www.usatoday.com/tech/science/2009-02-09-darwin-evolution_N.htm, last accessed 6/24/18.

http://en.wikipedia.org/wiki/Alfred_Russel_Wallace, last accessed 6/24/18.

www.guardian.co.uk/science/2008/jun/22/darwinbicentenary.evolution, last accessed 6/24/18.

RANGASWAMY SRINIVASAN–VISX PATENT DISPUTE

Summary: In 1983, Rangaswamy Srinivasan, the now retired IBM research scientist, collaborated with Stephen L. Trokel, an ophthalmologist at Columbia University to develop the technique that went on to become corrective eye surgery using lasers. In 1992, after their collaboration had ended for all intents and purposes, Trokel filed a patent, listing himself as the sole inventor. Srinivasan has not reportedly seen any of the financial benefits from his work. This is one of the most serious breaches of collaborative ethics in recent history.

The story: The ill-fated collaboration began in 1983 when Trokel convinced Srinivasan to help him (Trokel) develop a method for vision corrective surgery using an excimer laser. Not only did Srinivasan not know about a patent filed by Trokel in 1992, but there were other issues too, from the earlier days of their collaboration that would have caused a more cynical collaborator to walk away—something Srinivasan did not do. For example, a paper the pair coauthored was eventually found by Srinivasan to contain several errors, and a paper, which Srinivasan claims, was corrected by Trokel based on suggestions from the editors of the *American Journal of Ophthalmology*. Among the errors Srinivasan found were a misuse of at least one reference and a referral to papers by Srinivasan that were already in print as "in press" or "unpublished." This last point is important since the timing of a report can be used to establish inventorship. In fact, he claims that not a single article that was pointed in the direction of the phenomenon Srinivasan and his colleagues at IBM had described was found in the paper. The phenomenon was critical to the success of the laser eye surgery method. In 2000, an International Trade Commission (ITC) ruling regarding the patent in question declared Trokel's patent invalid, claiming Srinivasan should have been named coauthor. The ITC ruling goes on to state that Trokel knowingly and deliberately acted with deceit when submitting the patent. Other lawsuits have also been brought against Trokel and Visx, a company Trokel is a stakeholder in regarding patents also. Stockholders have sued the company, claiming that they had been misled by the fraudulent patents. Srinivasan has served as an expert witness in several lawsuits brought against Visx and Trokel, and, for his part, only hopes that someday he receives due credit for contribution to this work.

Questions to ponder:

1. Should Srinivasan have been more vigilant in the corrections Trokel did to the manuscript can authored together?
2. Did Trokel really do anything wrong or was it Srinivasan's fault for not pursuing things, especially the patent, more aggressively?
3. Are the shareholders of Visx owed anything? After all, the company has made millions of dollars, anyway.

SOURCE

Schulz, W. G. *Chemical and Engineering News*, **May 28, 2001**, 35–37.

SCHWARTZ AND MIRKIN

Summary: This is the case of Peter Schwartz, former postdoctoral research associate of Chard Mirkin's lab, and Chad Mirkin. This case covers several issues, including who has the right to publish and the ethical issues of publishing without the consent of all coauthors and failure to acknowledge the work of all coworkers. It is also highly important because it brings up other issues that do not quite fall into the category of any particular ethical violation. In particular, it involves the nature of the professional relationship between a research mentor/lab director and the junior scientists and how their (or one individual's) professional interests may come into conflict with the university's. There is a third individual involved with this story, Lydia Villa-Komaroff, the Vice President for research at Mirkin's institution whose comments are included below because they do carry weight and provide the university's response to this issue.

How did it start? Schwartz, after leaving Mirkin's lab, attempted to publish research that he performed in Mirkin's lab in the journal *Langmuir*. He tried to do so without acknowledging and without the consent of coworkers in Mirkin's lab and even and especially Mirkin himself.

Mirkin says: Upon learning that Schwartz submitted the article to *Langmuir*, and that the article was accepted, he wrote a letter to *Langmuir* with several objections. These objections included that Schwartz

1. Came into the lab with minimal expertise in the area of research.
2. Contributed to the group on one part of the project, then after some disagreements decided to leave the group and submit the entire research publication on his own.
3. Decided to submit the manuscript himself (without the consultation and without acknowledging coauthors).
4. Did not allow others in the lab to correct or verify the accuracy of the work.

The letter also claimed that

1. There is some question regarding the interpretation of the work.
2. The work has never been reproduced by Schwartz or the Mirkin's lab.

Schwartz says: Schwartz responded to several points in the following ways:

1. Schwartz received his PhD in 1998 from Princeton University, conducting research on the formation process of alkanethiol self-assembled monolayer on Au(III)—the very system behind his research with the Mirkin's lab that he was trying to publish and patent. He also worked with atomic force microscopy and scanning tunneling microscopy before joining the Mirkin group.
2–3. According to ACS author guidelines, he was the only significant contributor; however, he offered coauthorship to two members of the Mirkin's lab, an offer that was denied, with no one subsequently claiming authorship. He also claims that he offered coauthorship to Mirkin and that Mirkin

responded by forbidding Schwartz from contacting him, insisting that all communication go through the Vice President of Research, who subsequently failed to respond to Schwartz's emails and calls.

4. The research for the manuscript took place after he left the lab.
5. Finally, Schwartz addresses the reproducibility by claiming that he reproduced the results no less than 15 times.

Mirkin responds: In response to the points made by Schwartz, Mirkin claims the following:

1. Schwartz was one person in a group of 26 on a project well underway prior to his joining the lab.
2. Schwartz had no prior experience with nanoparticles and DNA and that his understanding of scanning probe microscopy was rudimentary compared to the group's.
3. It was made clear to Schwartz before his departure that no coworkers thought the work was publishable yet and that when it was, he would be offered coauthorship.
4. Schwartz presented group ideas, developed over several years, as his own.

Villa-Komaroff's role: For her role in the matter, Villa-Komaroff states that she stopped correspondence with Schwartz when he informed her that he was retaining legal counsel. She consequently informed him that all correspondence must now be through his attorney and the university's attorney(s). She has gone on to explain that the intellectual property actually belongs to Northwestern and that she had not released it, disallowing Schwartz from filing for a patent on the work.

There has been a resolution, for now, regarding this case. The work in question was accepted for publication by the journal with an addendum stating that the scientist who supervised the work takes issue with some of its content, its ownership, and aspects of how the work was carried out. You can find the addendum at the end of the journal article in question. Schwartz believes that his results cast doubt on the performance of a competing technique developed by the Mirkin group.

As of 2018, Peter Schwartz is now a professor in the physics department in the College of Sciences and Mathematics at Cal Poly, and Chard Mirkin remains at Northwestern. Both these statuses were found by performing a simple Internet search for both men.

Questions to ponder:

1. Did Schwartz behave unethically?
2. Did Mirkin and his group behave unethically?
3. Did the journal behave unethically?
4. Are the university's IP rules unethical?

SOURCES

Ritter, Steve *Chemical and Engineering News*, **2001**, June 18th, 40.

Chemical and Engineering News, **2001**, *July 30th,* 8–11-letters to the editor by
    Schwartz, Mirkin and Villa-Komaroff.
Adom, David *Nature*, **2001**, *412,* 669.
Schwartz, Peter, V. *Langmuir*, **2001**, *17,* 5971–5977.

COREY AND WOODWARD

Summary: In 1965, Robert Burns Woodward was awarded the Nobel Prize for his outstanding achievements in the art of organic synthesis. In 1981, Roald Hoffmann (along with Kenichi Fukui) was awarded the Nobel Prize for their theories, developed independently, concerning the course of chemical reactions. In 1990, Elias James Corey was awarded the Nobel Prize for his development of the theory of methodology of organic synthesis. Although the pedigree presented here has no real bearing on the case presented below, it does provide the perspective of the sheer magnitude and awesomeness of the characters involved. Corey claims that Woodward stole the idea (and Hoffman's Nobel) from him (Corey) during a conversation in Woodward's office that led to the Woodward–Hoffmann rules. Because Woodward died in 1979, there is no rigorous way to refute or support Corey's claim, though anecdotal evidence does both. Corey has repeatedly (both privately and publically) called upon Hoffmann to set the record straight. In 1961, 3 years before this putative conversation between Corey and Woodward took place, a Danish chemist at Leiden University, L. J. Oosterhoff, published an article in the journal *Tetrahedron*, first suggesting the very principles that later became the Woodward–Hoffmann rules. Woodward and Hoffman *correctly cite Oosterhoff's work*, which applied the theory that they later generalized to a very specific system.

Corey says: In Corey's defense, his story has been unwavering for years. Although this by no means constitutes proof of truth, it certainly lends credence to his claim. He claimed, during his Priestly Medal acceptance address in 2001 that he "suggested to my colleague, R. B. Woodward a simple explanation...conversations that provided the further development of the ideas into what became known as the Woodward-Hoffmann rules." Corey communicated with Hoffmann often through the years, writing him letters throughout the 1980s, imploring him to set the historical record straight. These letters are available at the Cornell University library (where as of this writing, Hoffmann still works). One of the keys to Corey's argument is that Woodward (according to Corey) originally opposed the proposal that Corey made. The very next day, Corey claims, Woodward used *his* (Corey's) explanation in a conversation with Professor Douglas Applequist, who was visiting. Woodward, according to Corey, presented the idea as his own, not crediting Corey, despite the latter's presence in the room (the conversation allegedly took place in Corey's office.) Applequist, who knew Corey's interests in the area, later went so far as to express a level of surprise that Corey was not one of the coauthors on the original paper by Woodward and Hoffmann.

Hoffmann says: First and foremost, Hoffman finds Corey's attack on Woodward to be unfair at this point due to the fact that the latter has been dead for decades. Although Hoffmann claims he does not recall a conversation with Corey during this time in which Corey told him (Hoffmann) of the large role he played in Woodward's development of the idea, he does admit that he asked Woodward if Corey should be included on the original paper proposing the idea. To this question, Woodward responded with a one-word answer: "NO." The respect, admiration, and unwritten code of junior and senior colleagues apparently compelled Hoffmann to not press the issue further.

Unlike Corey's account of the issues, Hoffmann's memory has not been as consistent, a fact he freely admits. This, however, does not mean he has been untruthful, nor does it mean he's been mistaken. Hoffmann has stated that, while he does not believe Corey's claims, if hard evidence could be presented, he would apologize and give credit to Corey for his seminal idea, setting the record straight in Corey's eyes. One final point that Hoffmann (and others) made in defense of Woodward is that it was not unusual for Woodward to ask questions of colleagues for which he already knew the answer.

L. J. Oosterhoff: Oosterhoff had no direct role in this controversy. His work published in 1961 and cited by Woodward and Hoffmann in their paper was the true first report of this theory. That they referenced it in their paper is proof-positive that they were aware of this prior work. To claim that Corey was unaware of this work by Oosterhoff would likely be a mistake; in all likelihood, he was indeed aware of it. Likewise, to claim that *both* Woodward and Corey were ignorant of this paper, and after discussion with Corey, Woodward did a literature search and found that Oosterhoff's paper is ludicrous. The year 1964 was a time of typewriters and telephones, not computers and the Internet. An overnight work session that would turn up such a paper and result in its thorough understanding is likely impossible, even for the best of the best, which Woodward arguably was. Furthermore, the likelihood that it was Corey who, with his suggestion, opened Woodward's eyes to this work is also low. If Corey's claim was instead that he *gave Woodward the Oosterhoff paper* as proof of his argument, the proof Hoffmann asks for may already be there. This argument not apparently being Corey's claim makes it unlikely that this was the way the events transpired.

Resolution: This is truly an immovable object vs. an unstoppable force. Corey is **not** going to convince Hoffmann that his version of events is true. The one man who can categorically refute or affirm Corey's claim has been dead for nearly 40 years. In all likelihood, this controversy will not be resolved any further.

Questions to ponder:

1. Should Corey have ever said anything?
2. What incentive would Corey have for waiting so long to voice these claims publically so late?
3. What, if anything, should Hoffmann do?
4. Should Hoffmann have pressed Woodward further?

SOURCES

www.boston.com/news/globe/health_science/articles/2005/03/01/whose_idea_was_it/, Last accessed, 6/24/18.

Chemical and Engineering News, letter to the editor, **4/28/03.**

Chemical and Engineering News, letter to the editor, **2/08/05.**

Chemical and Engineering News, letter to the editor, **11/29/04.**

CÓRDOVA

Scripps Research Institute and Stockholm University, 2003–2007.

Summary: This is the case of Armando Córdova, at the start of this story, a professor at Stockholm University. Córdova was found to be guilty of scientific misconduct for two out of the four charges brought against him, though investigators also commented that *many* other allegations and rumors of scientific misconduct could not be fully substantiated. The first two charges of which he was found guilty are discussed here. The most prevalent forms of scientific misconduct that we will encounter here are the failure to cite previous work and attempting to pass another's work off as one's own. Another form of scientific misconduct that we will encounter here is the failure to publish without the consent of all of the researchers. Many different individuals will be mentioned in this story in addition to Córdova: Donna G. Blackmond, whose tireless pursuit to protect her work from theft brought this story to light; Carlos F. Barbas III, a former mentor of Córdova at the Scripps Research Institute; Stefan Nordland, Dean of the Faculty of Science at Stockholm University; Olov Sterner and Torbjörn Frejd, independent investigators from Lund University; and Benjamin List of the Max Planck Institute, a victim of Córdova's theft. One of the key points, as we will soon see, is that even after he was "busted", Córdova continued, rather than ended these crooked tactics.

What happened? The first documented incident involving Córdova occurred in 2001 when, while a senior postdoc in the Barbas lab, he submitted a manuscript to the *Journal of the American Chemical Society* (where it was rejected) and then to *Tetrahedron Letters* (where it was accepted) without the consent of Barbas. Barbas was able to have the paper retracted, after discovering what happened by which time Córdova had been fired from the Barbas lab for reasons Barbas has not publically divulged. In early 2003, Barbas expressed concerns to Stockholm University that Córdova may once again attempt to submit the work for publication. Barbas's concerns were indeed well-founded as Córdova submitted the paper the following month to *Synlett*, where it was accepted with Córdova as the sole author. The Sterner–Frejd investigation ruled that this was "a clear case of unethical behavior." In 2003, *Synlett* published an addendum giving the history of the manuscript in question, including the comments from the Sterner–Frejd investigation that found Córdova guilty of unethical behavior with respect to this manuscript. The editors in the addendum state that "It is much to the regret of the Editors and the Publisher of *Synlett* that the paper has been published in *Synlett*."

Another incident in which the Sterner–Frejd investigation found Córdova guilty of scientific misconduct was the case involving Donna Blackmond of Imperial College in London. In late 2005, Blackmond delivered the Holger Erdtman Lecture at KTH, the Royal Institute of Technology in Stockholm, a lecture that Córdova attended. The work that Blackmond presented had been recently submitted to *Nature* for publication. An investigation determined that during the weeks after the lecture, Córdova's group studied a similar system and attempted to publish the results in *Chemistry-A European Journal*. This paper was accepted for publication and eventually read by Blackmond. In reading Córdova's paper, Blackmond recognized some of the concepts that were developed by her lab and that her work had not been cited. Blackmond went on to point out multiple weaknesses in the Córdova's paper and requested that he be required to publish a corrigendum. Córdova counterclaimed that Blackmond had actually gained insights from one of his previous papers that she had reviewed and that she failed to cite his work. Córdova's paper that he refers to as not being cited by Blackmond was rejected for publication, making it impossible to cite, and he later withdrew his allegations. The Sterner–Frejd investigation found evidence to support Blackmond's claims when they examined the Córdova lab notebooks. Finally, while Blackmond and Córdova settled the issue of a corrigendum to the *Chemistry—A European Journal* publication, Córdova submitted some of the same data and similar claims in the *Chemistry* paper to another journal, *Tetrahedron Letters*. This paper was accepted not only despite its similarity to already published and under dispute work but also despite not citing Blackmond's work properly. In this new paper, Blackmond's work was cited to be fair to Córdova, but it was buried deep in the paper and failed to give appropriate credit to Blackmond for being the first to report the observed behavior. As if these transgressions weren't enough, Benjamin List of the Max Planck Institute for Coal Research in Mülheim, Germany, claims that after speaking at a conference in Italy that Córdova attended, similar work to List's, published by Córdova appeared in *Tetrahedron Letters*.

The penalty that Córdova received was that he had to attend an ethics course and that all of his papers must be presented to his Dean before publication. Nordlund (Córdova's Dean) has stated that he feels that this penalty is sufficient, while also commenting that he "is very unhappy with and do(es) not support the behavior of Córdova" but going on to claim that "this is not the worst ethical behavior in science" in partial response to potentially firing Córdova for his scientific misconduct. Córdova, in his own defense, has steadfastly claimed that he didn't know that this sort of behavior was foul and that he was just following the mentoring of former advisors, a claim that Barbas has denied vociferously.

Resolution: To date, there have been no further officially publicized resolutions to this ordeal. Browsing the totallysynthetic.com blog, there have been unverified rumors that Córdova has been blacklisted by the ACS journals. Searching the ACS publication website for Córdova gives hits after 2006, the year that many of these incidents occurred, casting doubt on these rumors. With that being said, Córdova certainly does not seem to be doing too badly for himself, career wise. He is right now a professor in organic chemistry and a researcher at Mid Sweden University and Stockholm University while also serving as the head of chemistry at Mid Sweden University. Furthermore, in 2009, Córdova was selected for a professor chair in organic chemistry.

After a simple Internet search of Córdova turned up his professional profile page at Mid Sweden University, it is clear that any ban on publishing in ACS journals that may have existed (and remember these were rumors) has now been lifted as he has some publications in ACS journals in the past few years and also his work has been presented (by him or his workers) at ACS meetings.

Questions to ponder:

1. Was what Córdova did truly wrong?
2. Were the penalties levied on Córdova appropriate?
3. Should an entire family of journals blacklist individual authors the way the ACS appears to have done to Córdova?

4. Was Barbas wrong in trying to prevent Córdova's publication?

5. Did Blackmond overreact?

SOURCES

Chemical and Engineering News, **2007**, March 12th, 35–38.

http://Armandocardova.com, last accessed June 22nd, 2011.

http://totallysynthetic.com/blog/?p=322, found when searching goggle for Cordova and Blackmond, last accessed June 22nd, 2011.

Synlett **2003**, 2146.

www.miun.se/en/personnel/armandocordova/, last accessed 6/24/18.

WOODWARD AND QUININE

Summary: Quinine is a potent antimalarial drug (and for a while was the only recognized one). Its widespread need was perhaps greatest during World War II when access to the *cinchona* trees was blocked by the Japanese takeover of Java. Robert Burns Woodward and William von Eggers Doering answered that urgent call for a synthesis, though their work was never actually used to prepare this important drug. For reasons unrelated to the story that follows, Woodward and Doering never actually prepared quinine with this process.

What happened? Woodward and Doering titled their paper "The Total Synthesis of Quinine." In fact, their synthesis was *not* a total synthesis by today's standards; instead, it was what we call today a formal synthesis. It should be noted that the term formal synthesis was not widely used in 1944, the time of their publication, though it is certainly used today. These semantics have **no** relation to whether or not Woodward and Doering behaved unethically; they used terminology consistent with their era. It would therefore be a thorough waste of time to get hung up on their choice of words. What is the difference though? A total synthesis (in today's terms, anyway) is quite self-explanatory: it is the preparation of a target (usually a natural product) from commercially available starting materials. It is a *total* synthesis, something made from scratch. A formal synthesis (again, today's terms) is a synthesis that arrives at an intermediate that someone else has shown can be taken to the final product. The latter is what Woodward and Doering did. In the paper presenting their work, Woodward and Doering properly cited the work of Rabe, whose work in 1918 showed that the same intermediate they made could be taken to quinine. Several synthetic organic chemists since this time (among them, Gilbert Stork, who himself contributed the first stereoselective total synthesis of quinine in 2001) questioned whether or not Rabe's synthesis would have worked at all. Although Hoffman–LaRoche briefly explored Rabe's approach in the 1960s, they found it to need major alteration that made it impractical for them and published their own synthesis of quinine in the 1970s,[1] it was not until 2008 that researchers finally put the Woodward–

---

[1] This should not be interpreted as an indictment of the validity of Rabe's method. When pharmaceutical companies scale up production, they often have to greatly revise the syntheses.

Doering–Rabe method to a full test. Robert Williams and Aaron Smith at Colorado State in Fort Collins demonstrated that Rabe's synthesis worked reasonably well until the last step. They ultimately found that if one of the air-sensitive reagents in this step was exposed to air (conditions highly possible if not likely in Rabe's 1918 German lab), the troubling reaction worked quite well.

Resolution: Even Stork, who doubted Rabe's approach and argued strongly to have the record straight with Woodward and Doering's synthesis (all the while also admiring what the pair did), praised the work of Williams and Smith. Again, here, there may not be more to discuss in the way of misconduct. No misconduct was ever accused or even hinted at. Instead, what was subtly challenged is the norms of science. Is it "right" to take someone else's word for it? If we can't, the whole system falls apart because we are then forced to constantly reinvent the wheel. But, if we always do and never check, the self-correcting nature of science never operates. Often, any problems would have been found out if the Woodward–Doering intermediate and Rabe's synthesis were to be used commercially. By the time of the publication of Woodward and Doering's work (1944), the war was nearly over, easing the need to explore this route.

Questions to ponder:

1. Did Woodward and Doering act unethically by not completing the synthesis?
2. Did Woodward and Doering commit bad science by note completing the synthesis?
3. What were Stork's motives for being so vocal?
4. Should Hoffman–LaRoche have revealed the problems they encountered?
5. Assume for a moment Rabe fabricated the results. Should Woodward and Doering "pay" for that?
6. Is allowing a reagent to spoil in order to get the reaction work as Williams and Smith did bad science or amazing ingenuity in this case?

SOURCES

Rouhi, A. M. *Chemical and Engineering News*, **5/7/01**, 54–56.
Chemical and Engineering News, letter to the editor, **6/18/01**.
Chemical and Engineering News, letter to the editor, **8/31/01**.
Halford, B. *Chemical and Engineering News*, **2/26/07**, 47–50.
Halford, B. *Chemical and Engineering News*, **2/4/08**, 8.

DNA

Summary: The story that surrounds the elucidation of the structure of DNA is one of the most intriguing in modern science. The importance of the work that revealed the chemical structures of the molecule of life is impossible to overstate. It is, in fact, difficult to think of ten discoveries that supersede its importance (at least when you consider how many important breakthroughs were possible because of it.) What many people don't know, however, is that an enormous level of controversy surrounds this monumental discovery. Few controversies bring about as much of a tempest as this one does, in the right company. The cause for this? Rosalind Franklin and the way she and her work were (and to some still is) perceived to have been treated. To be fair, there is at least some evidence she may have gotten the short end of the stick. Unfortunately, the world will never know if Franklin would have shared the Nobel Prize with James Watson and Francis Crick (presumably at the expense of Maurice Wilkins); she died in 1958, before this trio was chosen for the award (1962). The award cannot be given to deceased individuals nor can it be awarded to more than three people. The argument can certainly be made that Wilkins contributed *significantly less* than Franklin did, and if she had lived, she would have gotten the award as a result. Awarding it to Wilkins then, perhaps, can actually be perceived as a nod to Franklin. Alternatively, Watson and Crick may have received one and Wilkins and Franklin another. We'll simply never know.

What happened? In 1951, James Watson arrived in Cambridge, in the Cavendish Laboratory, and was immediately teamed up with Francis Crick by Sir Lawrence Bragg. Thus began a friendship and work partnership that would quite literally change the world. At this early point, it is worth it to point out at least with regard to the elucidation of the structure of DNA, neither Watson nor Crick collected a single piece of experimental data themselves for this project. This undoubtedly is a partial source for the bile frequently spewed at them. However, this should take *nothing* away from the feat they accomplished. They put the pieces together (figuratively and literally) in ways nobody thought before them. Through attending seminars and meetings, they **legitimately** (at least at first) acquired data that led them eventually to the right answer. Their primary source of DNA X-ray data was Rosalind Franklin and Maurice Wilkins, who were coworkers in the loosest sense of the word at Kings College in London.

At this point, it should be mentioned that Wilkins and Franklin had a relationship that was several steps beyond strained. To blame for this, at least in part, was John Randall, head of the Biophysics Unit (Franklin and Wilkins's unit) at Kings College. Randall, while interviewing Franklin, explained to her that she would be working independently, while also intimating to Wilkins that Franklin would be working under him (Wilkins). It is no small wonder, with this in mind, why the two were more in opposition to one another than true colleagues should *ever* be. As for Randall's motive, one can only speculate now. It has been supposed that with Franklin and Wilkins' rivals, Randall would be more able to sweep in and share credit if either made a major discovery. Although some have wondered if the two would have **ever** truly coexisted working together even without Randall's stoking a fire, it is impossible not to speculate what could have happened. If they were able to work cooperatively, perhaps and truly just *perhaps*, they would have been able to beat Watson and Crick to the goal.

In 1952, Watson attended a seminar at King's by Franklin, an update on the last 6 months of her work. However, he was *sans* notebook as he often relied on a very keen memory. Much to especially Crick's dismay, Watson flubbed several details, including the number of water molecules (he was off by an entire order of magnitude.) What followed was an unmitigated disaster for Watson and Crick. They built a heinously incorrect model based upon Watson's recollection of Franklin's seminar. Making matters worse was their invitation to Wilkins and Franklin to come view their model for approval. Franklin immediately began pointing out flaws in the proposed structure, with the number of waters it contained being just one of her complaints. To say that she was less than impressed with the degree to which her results were misinterpreted would border on a comical understatement. Not surprisingly, word of this episode eventually reached Bragg, causing him significant embarrassment. His impatience with Crick already thin for unrelated reasons, this perceived besmirching caused him to forbid the pair from further work on DNA. At least part of his argument was that he considered it ungentlemanly and against the norms of British science to work on something another Brit was studying. Such toe-stepping was to be avoided in his opinion.[2] This could have been the end of Watson and Crick's contribution to the story of DNA and probably would have been if Lady Luck (or perhaps it was fate) hadn't intervened later in 1952, shortly after the episode with Franklin and Wilkins. Peter Pauling, son of Linus Pauling, one of the greatest chemists to ever live, joined the Cavendish. The importance of this was that Watson and Crick had long feared that the elder Pauling would eventually focus his considerable intellect on elucidating the structure of DNA, for he did related work. Later that year, a letter from father to son that the younger Pauling shared with his coworkers confirmed these fears. The great Linus Pauling was hot after the structure of the molecule of life and Watson and Crick effectively began an imaginary countdown to the day Pauling would announce that he'd done it. During this time, Franklin's X-ray pictures of crystallized DNA obtained from Wilkins were being collected with ever-increasing quality. Her photographs were even higher quality than anything Pauling had. One picture in particular clearly showed the helical nature of DNA. Despite this, she (Franklin) reported in an official in-house lab announcement that the helical structure was incorrect. It may be noteworthy that Wilkins was apparantly a supporter of the helical structure.

Early in 1953, the Pauling paper announcing the structure of DNA was released, and Watson and Crick were immediately *relieved*. His structure actually bore a remarkable resemblance to the failed structure Watson and Crick themselves proposed in 1952 just before his son joined the Cavendish. Pauling got it wrong. Convinced it was only a matter of time before he discovered and corrected his error, Watson and Crick resumed work (covertly at first) on DNA. The truth is, they never stopped thinking about it, just openly working on it. But, with this gaffe by Pauling, they decided they must act quickly and deliberately. Watson went to King's College. He was to meet with Wilkins but arrived early and decided to stop and see Franklin to discuss Pauling's error. At his harping on his and Crick's preference for the helical and maybe double helical structure, Franklin became enraged and if you believe Watson's account, frighteningly so. This does have some import here. Previously, it was mentioned that Franklin declared **no helix**, despite evidence to the contrary. Her reaction here suggests she really may have believed her interpretation to be true. If this is the case, she almost certainly allowed her friction with Wilkins to cloud her interpretation since Wilkins was also a supporter of the helical structure. Again,

---

[2] Remember this point later on!

one can only wonder what would have happened if these two worked *together*. If Wilkins and Franklin had been more teammates than adversaries, perhaps it would have been them to share the prize.[3] In any event, Watson was "saved" from Franklin's rage by the appearance of Wilkins and the two retired to his office. Here, Wilkins shared with Watson a copy of one of Franklin's X-ray photos. This photo contained remarkable clarity and, for Watson, absolutely confirmed the helix. Unbeknownst to Wilkins, Watson and Crick would soon obtain that same data (including some of his own (Wilkins's) data) from other means as well. One of Watson and Crick's colleagues, Max Perutz, was part of a committee appointed by the Medical Research Council, to which the Wilkins's and Franklin's results were sent, and Perutz shared these with Watson and Crick. Eventually, Watson and Crick were no longer able to hide their resumed pursuit from Bragg. Fortunately for them, Bragg and the elder Pauling were rivals. With Pauling now so obviously in very hot pursuit, the gentlemanly behavior Bragg was so beholden to earlier evaporated. Watson and Crick would now work on this monumental project with his full blessing. Moreover, they were permitted to employ the machine shop to build their model parts.

The next great epiphany came to Watson from the Chargaff rules that there are equal amounts of the bases, adenine and thymine, and equal amounts of the bases, cytosine and guanine in DNA. The clarification Jeremy Donohue provided about the structure of guanine and thymine in a discussion with Crick was the final piece of data they needed. All this data combined made the formal structure assignment academic for Watson and Crick. They assembled their model and the rest is history. Shortly thereafter, Wilkins and Franklin visited and approved the structure. The foursome agreed to publish all their research in series in the same issue of the journal *Nature*. A little while after that, Linus Pauling stopped in and agreed, Watson and Crick must have the correct structure.

Resolution: Only Watson remains alive from the main players in this story. He has claimed that if Franklin had lived, she would have shared the Nobel. Whether this would have been at Wilkins's expense (which would have probably been appropriate since, although Wilkins eventually collected his own data, Franklin's may have more directly lead to the elucidation) or on her own with another Nobel is unclear. Watson's opinion appears to be that it would have been at Wilkins's expense. Many of the people involved with this story acted like children at times. Watson claims that after the affair was over, Crick enjoyed a friendly relationship with Franklin, but at this point, he can only offer a secondhand account that cannot be corroborated.

Questions to ponder:

1. Did Bragg behave unethically in forbidding Watson and Crick from working on DNA?
2. Did Bragg behave unethically in changing his mind when allowing Watson and Crick to resume work?

---

[3] The intensity of the friction between these two is likely all three of their fault (Randall, Wilkins, and Franklin). The comments herein are not meant to vilify one person.

3. Did Wilkins behave unethically in sharing Franklin's data with Watson?
4. Did Randall behave unethically in setting Franklin and Wilkins against one another?
5. Did the younger Pauling behave unethically in sharing the letter from dad?
6. Did Watson and Crick behave unethically **not** collecting their own data?

SOURCES

White, M. Rivals: *Conflict as the Fuel in Science* Vintage Books **2003**, 231–273.
Watson, J.D. *The Double Helix: A Personal Account of the Discovery of the Structure of DNA*, Touchstone Books **2001**.

DAVID BALTIMORE AND THERESZA IMANISHI-KARI

Summary: This case is peculiar in the sense that the conclusions that were made from the fabricated data were eventually proven *true* by other studies. This has led some (including one of the authors of the study) to claim that it does not matter that data was fabricated. In this case, a postdoctoral researcher (O'Toole) brought to light results fabricated by her mentor (Imanishi-Kari) who was involved in a collaboration with the principal investigator and Nobel Laureate (Baltimore). This case is filled with issues of fabricated data and other forms of deception, sloppy science, trouble for the whistleblower, and even a full-blown **congressional** investigation.

O'Toole's side: Margot O'Toole, whose whistleblower case was discussed earlier, began working in the Imanishi-Kari lab in 1985. During her tenure there, part of her task was to replicate results that Imanish-Kari had previously observed. After many trials, O'Toole was unable to replicate these results when she worked on her own, causing Imanish-Kari to have several allegedly belligerent and emotional blowups. O'Toole was eventually assigned maintaining the colony to finish off her term in the lab. At one point, her new task required her to check breeding records to verify the pedigree of a particular mouse. What she found stunned her. First, the records appeared devoid of any organizational scheme. Second, and this is what ignited the controversy, O'Toole stumbled across data that agreed with the result *she* was obtaining; there was no indication that Imanishi-Kari's result had *ever* been observed. The actual results were, to O'Toole's figuring, misrepresented in a paper Imanishi-Kari published with Baltimore in the journal, *Cell*, which presented the results O'Toole could not replicate, though she did obtain the "correct" result (the one in the *Cell* paper) once, when assisted by a junior colleague of Imanishi-Kari and the progenitor of the work reported in *Cell*. O'Toole further asserted that the data she found in the books showed that there were defects in the study that were not acknowledged in *the published manuscript*.

Just 2 days after the discovery, O'Toole brought her concerns to a trusted researcher at Tufts University, Brigitte Huber, whom counseled O'Toole to bring the concern to Henry Wortis, who was supporting Imanishi-Kari's impending move to Tufts. As a result, Huber and Wortis, along with Robert Woodland, met with Imanishi-Kari. The trio concluded after the meeting (which O'Toole was **not** party to) that nothing would be done to correct the paper. Understandably, O'Toole was not satisfied with this response and brought her concern to Martin Flax, chair of the Department at Tufts. Flax's response, considering Imanishi-Kari was in the process of moving to his department, was baffling, informing O'Toole that the issue was the Massachusetts Institute of Technology's (MIT) problem. Huber and Wortis then met again with Imanishi-Kari, this time along with O'Toole. At this meeting, Imanishi-Kari produced two pages of data. During this time, O'Toole also met with Gene Brown, Dean of Science at MIT, who told O'Toole to formally charge fraud or drop the matter entirely. Brown contacted Herman Eisen, a collaborator on a grant with Imanishi-Kari that the funding agency dropped Imanishi-Kari from and requested he contact O'Toole. When they met, O'Toole presented the 17 pages she found to Eisen, whom, according to O'Toole, responded "that's fraud." Eisen has since claimed that he does not remember this and accuses O'Toole of being incoherent during this meeting.

Another meeting, arranged by Eisen, between Baltimore, Weaver, Imanishi-Kari, and O'Toole, took place a short time after. This new meeting produced no real changes—nothing was to be done.

Charles Maplethorpe: In summer 1986, the case took on a life of its own and out of the hands of O'Toole, at least partially. This is because, Charles Maplethorpe, a former student of Imanishi-Kari with a history of friction with her, brought the case out into the public eye. The result was a congressional hearing led by John Dingell, a Congressman from Michigan who was also pursuing other cases of fraud in science and has also done so since that time. At the hearing, Maplethorpe testified that he overheard a conversation between Imanishi-Kari and another researcher, David Weaver. During this conversation, Imanishi-Kari expressed she'd been having trouble with a particular reagent and was seeing the same result O'Toole would later observe. To Maplethorpe's knowledge, this issue was not resolved prior to the publication in *Cell*. Imanishi-Kari protests this account on the grounds that Maplethorpe is not a disinterested witness.

In response to the congressional hearing, Imanishi-Kari produced a loose-leaf notebook she claimed contained original data later organized into that form. At this point, O'Toole finally formally charged fraud, concluding, along with others, that some pages were faked. Meanwhile, a National Institutes of Health (NIH)-appointed panel commenced its own investigation, appointed by then director James Wyngaarden. The panel was chaired by Joseph Davie, and although it found serious flaws in the paper, it did not classify anything as misconduct. O'Toole insisted this finding was in err. Wyngaarden reopened the investigation, appointing a new panel and establishing an Office of Scientific Integrity.[4] Regarding the previously mentioned notebook, O'Toole's contention that it was faked was corroborated by the forensic services division of the Secret Service. Their examination concluded that the notebook could not have been produced at the same time as the work.

---

[4] This was the precursor to the Office of Research Integrity, demonstrating the long-lasting impact of this case.

Imanishi-Kari was eventually found guilty of two counts of fraud: the first was in the original paper in *Cell*, and the other was in the production of the notebook. The final report issued in 1994 by the Office of Scientific Integrity detailed 18 charges of scientific misconduct. The penalty was proposed to be a 10-year ban on applying for federal grants. This would have included not just National Institutes of Health (NIH) but the National Science Foundation (NSF), Department of Defense (DOD), Department of Energy (DOE), and others. ALL federal monies would be unavailable to Imanishi-Kari for 10 years, for sure a career-killing (and appropriate) punishment. She was also suspended from Tufts. She expectedly appealed to the Department Appeals Board of the Department of Human and Health Services, and the panel in the Office of Research Integrity ruled that the original case didn't prove its charges by a preponderance of evidence, and so Imanishi-Kari was again allowed to seek federal monies and her suspension at Tufts was lifted. The decision went on to criticize the Office of Research Integrity and, as discussed earlier, O'Toole as well.

Imanishi-Kari: Imanishi-Kari was not new to controversy when the case erupted. One involves her claims that she holds a master of science from the University of Kyoto. The Boston Globe's Tokyo Bureau investigated and found this to be false. Two members of the Laboratory of Developmental Biology at the University of Kyoto apparently once wrote a letter stating "...from these evidences it may be evaluated that her two years activities are equivalent in quality to complete two years master course at our University." The head officer of sciences at the University of Kyoto, Yoshio Mitabe, wrote in 1995 during an Office of Research Integrity investigation that "we could not find any official evidence that she was enrolled or employed in any position or grade at Kyoto University" going on to refer to the letter that Imanishi-Kari produced as a "private-level certification without administrative will." This becomes significant in that the University of Helsinki, where she received her PhD, mandated that foreign students with no more than a bachelor's degree would still be considered to be undergraduates. To gain admission into its doctoral program, the student must have already possessed a master's degree. The controversy about the master's degree, or lack thereof, could also have landed Imanish-Kari in hot water with federal agencies because it could be found to be a falsification of her qualifications and professional biography. It is also important to note that the National Cancer Institute recommended full funding for a joint grant—except for the work to be done by Imanishi-Kari's lab. This grant was to work on the very paper in dispute, discussed here. This blow came shortly before the paper was submitted to *Cell*, so in all likelihood, the grant would have funded the next step(s). Imanishi-Kari was also reputed to be a poor recordkeeper, someone who did not see experiments through and someone who sloppily did calculations in her head, sometimes making errors, off by an order of magnitude. The adage "Where there's smoke, there's fire" could apply well here.

Baltimore: This man, David Baltimore, is perhaps *the* reason this case took on the life that it did. Baltimore is a Nobel Laureate and thus a giant in his field. It is difficult to say that this case would have been afforded the same level of attention if it did not involve someone of his eminence. As the case progressed, Baltimore was appointed the President of Rockefeller University. By the time it mercifully ended, he was forced out of this position. It should be clear that at no point was Baltimore accused of fraud. Belligerence, at times ... yes, but never fraud. Although the reality is that **all** researchers are responsible for the work being reported, a good liar can get nearly anything past his or her collaborators. This appears to be what Imanishi-Kari did to Baltimore, though it can certainly be argued that Baltimore's fame is why he was not also accused of fraud. One particularly troubling comment that Baltimore made however bears mention. Baltimore contends that since independent studies have more or less led to the same conclusions, it is not a big deal since the fraud clearly did not hold the science back. He also has allegedly wavered on his stance regarding the importance of the fabricated work's conclusions, on the one hand being reputed as implying that the contested conclusions are the whole point and then also claiming that they are minimal in importance.

The public perception: The perception of this case, in particular, within the scientific community was curious. At one point, during the congressional hearings, scientists were called upon by Philip Sharp, a MIT scientist and a friend of Baltimore, to write their congresspersons and submit op-ed letters to their local newspapers railing against the hearings, perceived by some to be nothing more than a witch hunt; an intrusion into science by lawmakers. This is an unfair characterization of the zeal with which Dingell pursued this case and others during that time. For sure, Dingell appeared incorrigible. However, the federal government *should* investigate with rigor accusations of fraud in cases that involve federally funded research. For certain, any congressional investigation **must** heavily rely on the opinion of research scientists. With this qualifier, the decision can be made regarding official penalties with an informed mind. This says nothing, however, nor does it have any influence over the unofficial penalties that ultimately are levied upon scientists who are perceived to commit fraud. Often, this stigmatization follows them forever, and since the author of a grant or manuscript is rarely anonymous,[5] it inevitably affects the review process, unleashing the self-policing power of science.

Summary: Out of necessity, many of the finer details were left out here. The interested reader is encouraged to read the account by Horace Freelance Judson, who includes firsthand observations and conversations in his narration. Enough information is hopefully included here to stimulate a provocative discussion of what was right and what was wrong here. When all that occurred is analyzed, one cannot help but conclude that Imanishi-Kari got away with it. Baltimore, if guilty of anything, is guilty of allowing her to do so. O'Toole, the steadfast defender of what was widely agreed as being the truth, displayed much of what is good and right about science. The Dingell investigation demonstrated both why Congress should get involved and with what was later perceived to be a bungled investigation, and why it should not.

Simple Internet searches via Google for each of the main players turned up that Baltimore is currently on the board of sponsors for the Bulletin of the Atomic Sciences. Imanishi-Kari remains at the Sackler School of Graduate Biomedical Sciences at Tufts, where she holds the rank of associate professor of immunology, according to her faculty profile page, found with a simple Internet search of her name

---

[5] Such reviews are called double blind.

and Tufts. Margot O'Toole's LinkedIn page claims she is a retired translational medicine scientist, now active in educational reform in the greater Boston area.

Questions to ponder:

1. Should Baltimore have felt more heat?
2. What could Baltimore have done differently?
3. Did O'Toole do anything wrong?
4. Did Maplethorpe do anything wrong?
5. Was Sharp's call to scientists appropriate or inappropriate?

SOURCES

Judson, H. F. *The Great Betrayal: Fraud in Science* Harcourt, Inc. **2004**, 191–243.
https://thebulletin.org/board-sponsors-0

JOHN FENN-YALE PATENT DISPUTE

Summary: John B. Fenn shared the Nobel Prize in Chemistry in 2002 with Koichi Tanaka and Kurt Wurthrich for his work in the development of electrospray ionization for the analysis of large molecules. Fenn filed patents on his work, and in 2003, courts ruled in Yale's favor that the patent should be Yale's property.

The story: Fenn, in the late 1980s, developed the electrospray ionization mass spectrometry method, allowing the mass spectrometry of large molecules to be measured reliably for the first time. In 1992, Fenn applied for and was awarded a patent on the method, with himself as the assignee, meaning he would be the principal earner of any monies gained. This is very atypical for scientists, especially at academic institutions. He then licensed the patent to a company he cofounded, Analytica of Brandford, Conn, who then sub-licensed rights to instrument makers. This last part is a typical protocol. By this time, Fenn had relocated from Yale to Virginia Commonwealth University, having been forced into retirement by Yale.

When Yale discovered this patent, it claimed rights to the patent, asking Fenn to reassign the patents to the University, a request Fenn refused. Fenn subsequently sued Yale. Yale filed a counterclaim, claiming that Fenn "misrepresented the importance and commercial viability of the invention...actively discouraged Yale from preparing and filing a patent application," going on to claim that Fenn inappropriately filed the patent himself, neglecting to notify Yale or the National Institutes of Health (NIH), which provided funding for the work. Fenn defended himself, claiming that Yale did not pursue a patent because it did not view it to have sufficient commercial interest, changing course when the University realized it was earning a substantial income for Fenn. He went on to point out that earlier electrospray patents **are** assigned to Yale and have been valuable to the University.

If, as Yale claimed, Fenn undersold the potential value of the patent to the university, this would certainly account for Yale's decision to not pursue a patent. In 2005, a court gave its final decision, ordering Fenn to pay nearly $1 million in combined misdirected royalties and legal fees, and that the patent be transferred to Yale, convicting Fenn of civil theft. In December 2010, Fenn, aged 93, passed away.

Questions to ponder:

1. If Fenn did undersell the value of the patent to Yale, did he commit misconduct?
2. If Yale exaggerated Fenn's initial claims regarding the commercial value of the work, should they have been awarded the patents?
3. Is it at all unethical that the University receives a significant cut of patent royalties?

SOURCES

http://en.wikipedia.org/wiki/John_Fenn_(chemist), last checked 9/22/11.
www.washingtonpost.com/wp-dyn/content/article/2010/12/11/AR20101 21102387.html, last checked 9/22/11.
Broman, S. *Chemical & Engineering News*, **February 21, 2005**, 11.

VIOXX

Summary: In 2004, pharmaceutical giant Merck voluntarily removed its painkiller Vioxx from the market. It was found during the course of a 3-year study on some other activity of the drug (after the Food and Drug Administration (FDA) approved it as a painkiller) that the drug placed participants at higher risk of cardiovascular events. If Merck's records are correct, and it's important to note that there is little reason to doubt they are, the side effects were not known when the FDA approved the drug in 1999. It is important to know that all the data Merck would have collected during clinical trials would have been shared with the FDA in order to gain approval, meaning Merck would have had to falsify the results sent to the FDA to get Vioxx accepted amid serious side effects. It is also in Merck's favor that the withdrawal was voluntary. They found that there was a problem and took the appropriate steps to rectify it. As a result (at least in part) of the issues Merck uncovered with Vioxx, one of the Merck's competitors, Pfizer, initiated a large study to investigate its own drug, Celebrex, which is a member of the same general family of drugs as Vioxx.

It is certainly not unheard of for drugs to be used and eventually approved for new treatments. Therefore, Merck exploring this for Vioxx was not at all unethical and, in fact, is quite typical. A perfect example of this is aspirin. Most of us grew up using aspirin as a painkiller. Today, it is also approved for treatment during a heart attack and after suffering a stroke. For years, it was informally used for these additional purposes. Now, aspirin has been around a very long time, much longer than Vioxx, so its safety record was already well-established, making it easier on aspirin. Pharmaceutical companies often redeploy their drugs for additional use. Sometimes, some of the uses are unofficial, occasionally staying that way. Others eventually gain multiple approvals.

This case is different than the others here in **Part B**. This is deliberate: it has been included specifically to open up the discussion about drugs being used for multiple purposes, including what dangers may be associated with this. When they are not approved uses, they are referred to as "off-label" uses. A list of drugs, their approved use(s), and the off-label uses are provided below.

Name of Drug	Official Use	Off-Label Use
Methotrexate	Choriocarcinoma	Unruptured ectopic pregnancy
Sertraline	Antidepressant	Premature ejaculation in men
Adderal and Ritalin	ADD in children	ADD in adults[a]
Gabapontin	Seizures and post neuralgia in adults	Bipolar disorder, essential tremor, hot flashes, migraine prophylaxis, neuropathic pain syndrome, phantom limb syndrome, and restless leg syndrome
Viagra[b]	Erectile dysfunction	Pulmonary hypertension

[a] Yes, it matters that they're different!

[b] An interesting note about Viagra: Viagra was initially intended to be a heart medication. During clinical trials, a side effect was reported by a large number of men in the study. Pfizer quickly realized that this side effect, if it is a *bona fide* result, would make them much more money than the original use, and so they revised their application and likely patents to reflect this. Millions of dollars later, the rest is history.

Questions to ponder:

1. Should it be allowed that drugs are used for any purpose other than the one they went through clinical trials for?

Vioxx update! Start-up Tremeau Pharmaceuticals is looking to bring embattled Merck drug Vioxx back to the market. In late 2017, the FDA granted Tremeau orphan drug designation for its intentions with the once-blockbuster painkiller. Tremeau hopes to market the drug for patients with both extreme joint pain and hemophilia since many other painkillers cannot be used by such patients. The company is now raising the funds to run clinical trials, and in February 2018, it announced that just over $5 million had been raised from investors so far. Troubling to some is that although (if approved for reentry to the market of course) the maker cannot advertise a use not approved, doctors can prescribe the drug for any manner of pain they see fit.

Importantly, two other options for these target patients currently exist. One is high doses of Tylenol. Tylenol, at very high doses, is toxic to the liver. Another option is opioids, which carry their own set of complications (addiction, overdose, and death). Tremeau is still reportedly deciding on a drug name. Vioxx is no longer protected by trademark, so they could (legally) use it, but this would no doubt scare as many users away from it as it would attract. If their current press is any indication, they appear to be moving away from the Vioxx name. They appear to be using the generic name rofecoxib and the code name TRM-201. Notably, their press release to prnewsire.com in February 2018 did not mention Vioxx by name once, though the name did appear in the citations in a title of a cited work.

It is worth commenting that in preparing this update, a website facts-are-facts.com was found to have a story on this topic. Others also have commented on this potential return to the market by Vioxx. In an editorial to the *British Medical Journal*, Drs. Joseph Ross and Harlem Krumholz had some recommendations, should Vioxx return to the market rebranded. They also comment in their editorial that at least at the time of initial approval, Merck reported some of its clinical trial data in a way that misrepresented results and may have obscured the risks. The authors also mention the likelihood of more widespread use than the license grants (which is perfectly legal; the FDA grants that something can be sold on the market and doctors are free to prescribe it for whatever they see fit, though few do so wantonly, and the manufacturer is not allowed to advertise any use but what it is approved for). They offer the following suggestions as safeguards to protect the public:

1. Routine collection and independent arbitration of safety data.
2. Consistent with the FDA Amendments Act, all clinical trials should be registered with clinicaltrials.gov before initiation and results should be reported within 12 months of completion.
3. The FDA should also monitor off-label use if approval is granted, proposing that if a threshold of total use (they offer 20%) is exceeded, the FDA would require that the manufacturer conduct studies to evaluate safety.
4. Insurance payers should adopt formulary management strategies that restrict use when alternative therapies exist.

This is a story worth watching very carefully in future years. By the time this book is in print, it will likely be changed! This case may well be precedent-setting for other "troubled" drugs.

Additional questions to ponder:

1. Should any entity be permitted to bring a drug, removed from the market due to safety, back to the market for any reason?
2. Is the potential to be an alternative to opioids enough of a benefit to warrant Vioxx's reinstatement?
3. Would the proposed restrictions be enough? If not, assume it is granted reentry into the market and suggest additional measures.

SOURCES

Marx, V. *Chemical and Engineering News*, **10/4/04**, 8.
Marx, V. *Chemical and Engineering News*, **10/25/04**, 15.
Mullin, R. *Chemical and Engineering News*, **11/15/04**, 7.
Chemical and Engineering News, letter to the editor, **1/3/05**.
www.prnewswire.com/news-releases/tremeau-completes-52-million-equity-raise-to-fund-non-opioid-pain-treatments-for-rare-diseases-300599188.html, last accessed 6/24/18.
British Medical Journal, 2018, 360, 1242.
Statnews.com, startup seeking to resurrect Vioxx, arthritis drug pulled from the market, AP 11/21/17, Steven Senne.

AUTISM AND THE MEASLES, MUMPS, AND RUBELLA VACCINE

Special note: I have two family members whom are on the autism spectrum. I've had multiple students on the autism spectrum. I, in no way, take this disorder and the stresses it puts on people and their families lightly. The direct causes of autism are still largely unknown, though vaccines have been a popular target. It is important to understand that there are still no well-designed and controlled studies I am aware of that have shown a causal link between the two. When something bad happens, it is natural to look for a scapegoat, for someone to blame, and for someone "to pay for what happened to me." Alas, sometimes, there is no one. Sometimes, it is just life and we need to make the best of it. Autism has a wide range of severity. My cousins are lucky to be mostly normal; they are quite like their "normal" cousins. Part of why may be genetic, and part of why may be that they had access to amazing social workers during very formative years. We'll probably never know.

The case

Andrew Wakefield was a medical doctor in the United Kingdom at that time. He has since been stripped off his rights to practice medicine over the row that the now retracted study detailing a link between the MMR vaccine and autism. The paper was retracted after a council found that Wakefield and at least one of his coauthors committed various forms of scientific fraud. The findings included the following:

1. The patients in the study were not randomly chosen and rather, in some cases, were even referred to the study.
2. The tests, performed on children who were sick, in some cases, mysteriously sick,[6] were done without ethical oversight but with consent from their very stressed and therefore vulnerable parents.
3. Wakefield was being paid, hourly, by a lawyer who was spearheading a lawsuit against vaccines. The same lawyer at least partially funded the research.
4. Wakefield held a patent on a vaccine regimen that split the trivalent vaccine into three individual vaccines, something he advocated for in the now retracted paper.
5. Wakefield neglected (or failed) to declare points 3 and 4 in the paper.
6. More than half of the authors of the paper disavowed themselves of the conclusions drawn by the paper.
7. It appears that some of the patients were removed from the study.
8. Parents of at least one of the children in the study claim that some of the data about their child was changed.

By the end of a long and extensive hearing, the *Lancet* (the journal where the work was published) fully retracted the paper. Although it is highly likely that the truth would have won out in the end, I believe it is unlikely that it would have happened on the timeline it did without the efforts of Brian Deer, a British investigative reporter. Deer authored a series of articles calling out Wakefield and his coworkers for their actions. His hard work almost certainly at the very least hastened the inquiry and perhaps is the direct cause of it.

The counterpoints

Naturally, if vaccine use were to be halted, a great deal of money would be at stake. Even if all currently deployed vaccines had to be reformulated, the financial

---

[6] I don't mean to insinuate that Wakefield made them sick, which has never been even hinted. What I mean is that they were sick, and their parents didn't know what was wrong.

toll would be staggering since the new formulations would have to be rigorously tested for safety. For sure, there will always be claims that the pharmaceutical industry has a vested interest in suppressing the "results" that clearly prove a link between autism and their products. There nevertheless is no credible nor to my knowledge even dubious evidence of a cover-up by the pharmaceutical industry. As a precautionary measure, the CDC and the American Academy of Pediatrics asked vaccine manufacturers to remove thiomersal from vaccines.

Points to ponder:

1. Imagine a day 25 years from now when a controlled and well-designed study proves that this vaccine causes Autism in a set of genetically susceptible individuals. Would this exonerate Wakefield and his coworkers? Why or why not?
2. What sorts of legal ramifications should someone face when found guilty of what Wakefield has been found guilty of?
3. If 1,000 children were to die because of a measles outbreak, facilitated by parents opting out of vaccines because of the now retracted paper, should Wakefield face manslaughter charges?

SOURCES

Brian Deer, *British Medical Journal*, three-part series
1. How the case against the MMR vaccine was fixed, www.bmj.com/content/342/bmj.c5347, last checked 6/24/18.
2. How the vaccine crisis was always meant to make money, www.bmj.com/content/342/bmj.c5258, last checked 6/24/18.
3. The Lancet's 2 days to bury bad news, www.bmj.com/content/342/bmj.c7001, last checked 6/24/18.

Rachel Sheremeta Pepling, Chemical and Engineering News, **2008**, December 15th, 34. "When controversy shouldn't exist".

https://en.wikipedia.org/wiki/Thiomersal, last accessed 6/24/18.

ITALIAN EARTHQUAKE SCIENTISTS

In 2012, an Italian judge made some waves and sentenced seven Italian natural disaster experts to 6 years of jail time each for giving false assurances before an earthquake that hit the city of L'Aquilla in 2009. Although six of the seven were acquitted on appeal, the seventh was not, though the sentence was reduced. The issue was a major earthquake that ultimately struck the area, even after a series of much smaller tremors over a period of several months. Only Bernardo De Bernardinis, at the time, Head of the Italian Civil Protection Department, was found guilty. His "crime" was that he failed to communicate the risk, which was his job. The judge in the case took pains to emphasize that it wasn't that he was guilty of failing to predict the earthquake.

At the time, a conference led to the impression that because of the smaller tremors, the energy that would be a larger quake was being released more slowly, and thus would not result in a larger quake. Some evidence, notably radon emissions, seemed to suggest some that a larger quake was still coming. This, however, was not fully communicated to the public, and the result was that many people were (or at least felt) unprepared for the big quake that ultimately did happen. It was argued, successfully in court, that had they been better warned, less damage and loss would have occurred since they would have behaved differently. This argument seems strange, at least to me, since it would take years, if not decades to prepare buildings for an earthquake.

Points to ponder:

1. Should the scientists have done better to convey the possible damages?
2. If you live in an earthquake-prone area, does basic personal responsibility and common sense demand that you always be ready for a larger earthquake?

SOURCES

The Guardian, "Italian Scientists convicted for 'false assurances' before earthquake", accessed 3/18/14.

CNN, "Italian Scientists resign over L'Aquila quake verdicts", accessed 10/23/12.

Science magazine, "Why Italian earthquake scientists were exonerated".

THE RED RIVER FLOODS

In 1997, the Red River of the North in the United States and Canada had a major flood event. Each year, the weather services in the area make predictions about how high the river will rise. They do so because the snow melt, combined with spring rains, make, for flooding to some degree nearly every year. With the complex models, they are able to make pretty accurate predictions and forecasts about the flooding, allowing the communities to prepare in ways such as building levies and dikes.

In 1997, the predicted levels fell short of the actual flood levels by several feet. The result was disastrous. The weather service separates their predictions into outlooks and forecasts. The forecasts are far more current than the outlooks, which are a sort of far-looking prediction. In this particular year, heavy snow made clear that this would be a bad year, and the outlook was posted earlier than usual and was higher than usual. As the spring melt intensified, the actual forecast crept higher and higher, eventually not only exceeding the original outlook but also the extra preparation on top of the outlook that had been put in place. The reason for it is almost a comedy of errors of sorts. A late blizzard added a significant snow to the melt, and a sudden late cold snap created ice dams in the river. Both of these factors were unexpected. Unseasonably warm temperatures then followed, which kept the water from refreezing each night, causing the ice and snow to melt faster than anticipated from historical data. All of the flood measures in place eventually failed and communities were nearly completely washed away.

Points to ponder:

1. Is it really anyone's fault when things that are basically impossible to foresee happen?

2. Is it really a surprise, in light of a story like this, that weather forecasts tend to "overblow" storm total predictions?
3. Is it really responsible to keep crying wolf with regard to weather predictions?
4. Should forecasters see punishment such as fines or jail time for forecast errors?

SOURCES

Morss, R. E.; Wahl, E. *Environmental Hazards*, **2007**, *7*, 342–352.

ANIMALS LOST DURING STORMS OR OTHER ACTS OF NATURE

In October 2012, the weather event dubbed "Superstorm Sandy" struck the downstate New York, New Jersey, and Connecticut areas. The damage from the former hurricane Sandy was staggering. Included among the millions of dollars in damages and human lives lost were damages to truly irreplaceable research samples and animals. The loss of the samples was largely due to the long-term power outages that resulted from the storms. The animals, on the other hand, were largely lost due to the floodwaters; they drowned.

At New York University (NYU), in Manhattan, the floodwaters inundated the building the animals were housed in. Although the buildings were built to code for the flood zone, nothing could have prepared them for this historic storm. As the floodwaters rose, the pumps in the buildings were unable to keep up with the rising water, and the result was thousands of mice and hundreds of rats perished in their cells. NYU noted at the time that the vast majority of their animals did not die during this storm, and in fact, the number of animals that perished was ultimately found to be lower than initially thought due to the actions of the lab manager ordering the cages to be put higher up in the rooms ahead of the storm. NYU also noted that the building had been built assuming a flood 20% worse than the worst historical flood of the past century at the time. Unfortunately, this storm obliterated historical records, and despite all continuous efforts, rescue attempts were unsuccessful. At least at the time, the Guidelines for the Care and Use of Laboratory Animals did not explicitly prohibit that animals are stored in the basement of a building. Disaster plans needed to be in place, and all indications are that such plans were in place but failed due to the unprecedented nature of the storm. In fact, the basement is, at times, an ideal place for animal storage. First, it allows the researchers to better control the day–night cycle of the animal, a clear research benefit in some cases. Second, in California and other earthquake-prone areas, it is safer. Third, disease-causing microbes carried by some animals are less likely to enter the air circulation system. More skeptically, it keeps them out of sight, and out of sight is out of mind. Yet more skeptically, it keeps things secretive and out of the watchful eyes of PETA and other organizations.

NYU is not alone in such loss. Hurricane Allison in 2001, Houston, and Hurricane Katrina in 2005, New Orleans, both ended the lives of thousands of lab animals, cumulatively, in excess of 20,000. It is clear that better planning is necessary since this seems to happen every handful of years or so.[7] Some better planning does seem to be in the works. For example, NYU is planning to move animals to higher floors, and the University of Texas is making a move to a new building with animals on higher floors as well.

Points to ponder:

1. Since the buildings were in line with the building code for their area, is there really a "fault" involved?
2. Since all the ethical oversights seem to have been adhered to, even if with the benefit of hindsight, the researchers should have known better than to house animals in the basement, should they face sanctions?
3. Should new buildings all have dedicated elevated spaces for animal research hereafter?
4. Should older buildings be "grandfathered" in? For example, should an animal lab in the basement of a building in a zone where there is virtually zero chance of flooding be forced to relocate, even if that means spending tens of millions of dollars on a new building?

SOURCES

Yahoo! MAKTOOB News article, "NY University faces growing criticism after Sandy kills lab mice" last accessed 5/8/13.

NPR article, "A tale of mice and medical research whipped out by a superstorm" last accessed 6/3/16.

TRANSFER OF A PATENT TO AN INDIAN TRIBE TO AVOID ITS INVALIDATION

This case is the result of a plan by Texan intellectual property lawyer, Michael Shore. In 2012, Congress passed the America Invents Act, which streamlined the procedures for the adjudication of patent challenges by the United States Patent and Trademark Office. Stone noticed, among other things, that by this new act, that the law was written in a way that did not apply to foreign countries and Native American nations. He further figured that this would make patents held by such sovereign entities worth significantly more than the same patent held by institutions subject to the new procedure. He then set out to find a suitable sovereign and eventually partnered with the St. Regis Mohawk Tribe in New York State, near the border with Canada. He then searched for a non-sovereign patent holder and found Allergan, whose patents for its major drug Restasis were being challenged by two companies, Mylan and Teva, both of whom were hoping to produce generic versions of the drug. The patents on this drug gave Allergan over 1.5 billion dollars per year monopoly in the area. Shore brokered a deal where Allergan would transfer the patent to the St. Regis Mohawk Tribe who in turn would lease it back to Allergan who would pay the tribe $15 million per year in addition to $13.75 million up front. Predictably, an uproar ensued.

---

[7] I guess we're kind of due to have it happen again.

In October 2017, a federal judge in Texas ruled that some of Allergan's Restasis patents were invalid. Allergan indicated it would appeal. The judge also had critical words for Shore's scheme, saying "Sovereign immunity should not be treated as a monetizable commodity." Some members of Congress, led by Senator Claire McCaskill, agreed even to the point of calling for the revoking of Native American sovereign immunity in patent claims cases, a response Shore claims would merely penalize Native Americans and not close the loophole that inspired the scheme to begin with. He goes on to say that if the option of Native American sovereign nations is closed, he would pursue state universities, going so far as to say he's in talks with several underfunded state-run historically black colleges and universities, who would similarly benefit in Shore's estimation.

The final blow could have been dealt in February 2018. The Patent Trial and Appeal Board, a court run by the United States Patent and Trademark Office, ruled that tribal immunity does not apply to patent review proceedings. The tribunal went on to say that (presumably due to the very large fees Allergan paid to the tribe) Allergan retained an ownership interest in the patents, indicating that this meant the proceedings could proceed sans the tribe's participation. However, a month later a federal Circuit Court granted a stay of the challenge to the patent. The inter-parties review had not started by the time this book went to press, but it was unanimously voted to move forward in July, 2018 (https://arstechnica.com/tech-policy/2018/07/court-native-american-tribe-cant-be-a-sovereign-shield-during-patent-review/).

Meanwhile, a bipartisan group of senators, led by Senator Claire McCaskill, introduced legislation that would prevent companies from "renting" sovereign immunity in the way that Allergan tried. The bill, called the Preserving Access to Cost Effective Drugs Act, had not been voted on when this book went to press, but it had been brought to the Senate floor.

Questions to ponder:

1. Should such a tactic to protect a patent, trademark, or any other form of intellectual property be allowed?
2. Was Shore unethical for trying to exploit a loophole he found?

SOURCES

The New Yorker, the Financial Page, "Why is Allergan Partnering with the St. Regis Mohawk Tribe? Inside the bizarre world of patent law. Adam Davidson, 11/20/2017.

Reuters, Health News, U.S. Patent Court Deals Setback to Allergan's Restasis Strategy, Jan Wolfe, 2/26/18.

FiercePharma, Allergan's much-maligned tribal licensing deal wins reprieve at appeals court, Eric Sagonowsky, 3/29/18.

www.bna.com/eyes-senate-bill-n57982089831/, last accessed 6/24/18.

Index

DNA, 13, 35, 119–123
DOE, *see* Department of Energy
Double-blind reviews, 126
Dow AgroSciences, 40
Duke Associate Professor, 37

E

Economic Espionage Act (1996), 40
Electromagnetic field (EMF), 82
E-mail
    Schwartz and Mirkin case, 113
    subpoenaed, 25
Environmental enrichment, 63
Environment and science, 85–87
Ethical violations, ways of committing, 4
Ethics
    bad science *vs.*, 30–32
    course, 117
    definition of, 3
    oversight committees, 42

F

Fabrication of data
    example, 6
    grant application, 41
    journals, 41
    peer review, 50
    penalty for, 35
    prevention of, 102
    reasons, 6
    violations, repercussions, 5
Facebook, 18
Fact-checking, level of, 51
Failure to acknowledge all researchers who
            performed the work, 18–20
    description, 18–20
    how it is caught, 20
    prevention of, 104–105
    why it happens, 20
False credit, 11
Fast Track process, 73–74
FDA, *see* U. S. Food and Drug Administration
Federal Economic Espionage Act of 1996, 40
Federal government, intervention by, 89–93
    blogfest, 90
    constitutional rights, 89
    embargoes, 89
    First Amendment, 92
    funding of controversial research, 92
    intellectual property, 91
    lawsuit, 89, 92
    public relations problem, 90
    publisher restriction, 89
    stem cell research, 92, 97

summary, 93
First Amendment, 92–93
Fracking
    company, 25
    controversy of, 57
    and pollution, 82
Frank's case, 52
Funding controversial research, issue of, 92

G

Geneva declaration, 56
German Physicians, 55
GlaskoSmithKline (GSK), 25, 90
Global terrorism, 95
Government, *see* Federal government,
            intervention by
Grammar snob, 48
Grant(s)
    application
        conflict of interest issues, 27
        fabricated data on, 41
    National Science Foundation, 16, 45
    paying back percentage of, 101
    as proposed projects, 14
    work stolen from, 12

H

Harvard, New England Primate Research
            Center, 43
Health and Human Services (HHS), 95–96
Helsinki declaration, 56
H-index, 11
Hippocratic maxim, 58
HIV, *see* Human immunodeficiency virus
HIV vaccine, 80–81
Human and animal subjects, 42–44
Human/animal subjects in research, responsible
            conduct, 55–67
    Animal Welfare Act, 42, 43, 62
    antibiotics, strains resistant to, 66
    anti-HIV drug, 61
    assessment of risks and benefits, 60
    drug company profit, 66
    drug testing, 66
    drug toxicity, 66
    ethical principles, 57–59
        beneficence, 58
        informed consent, 58
        justice, 58–59
        respect for persons, 57–58
    FDA rules, 57
    Hippocratic maxim, 58
    humans as guinea pigs, 61
    informed consent, 59–60

Printed in the United States
by Baker & Taylor Publisher Services